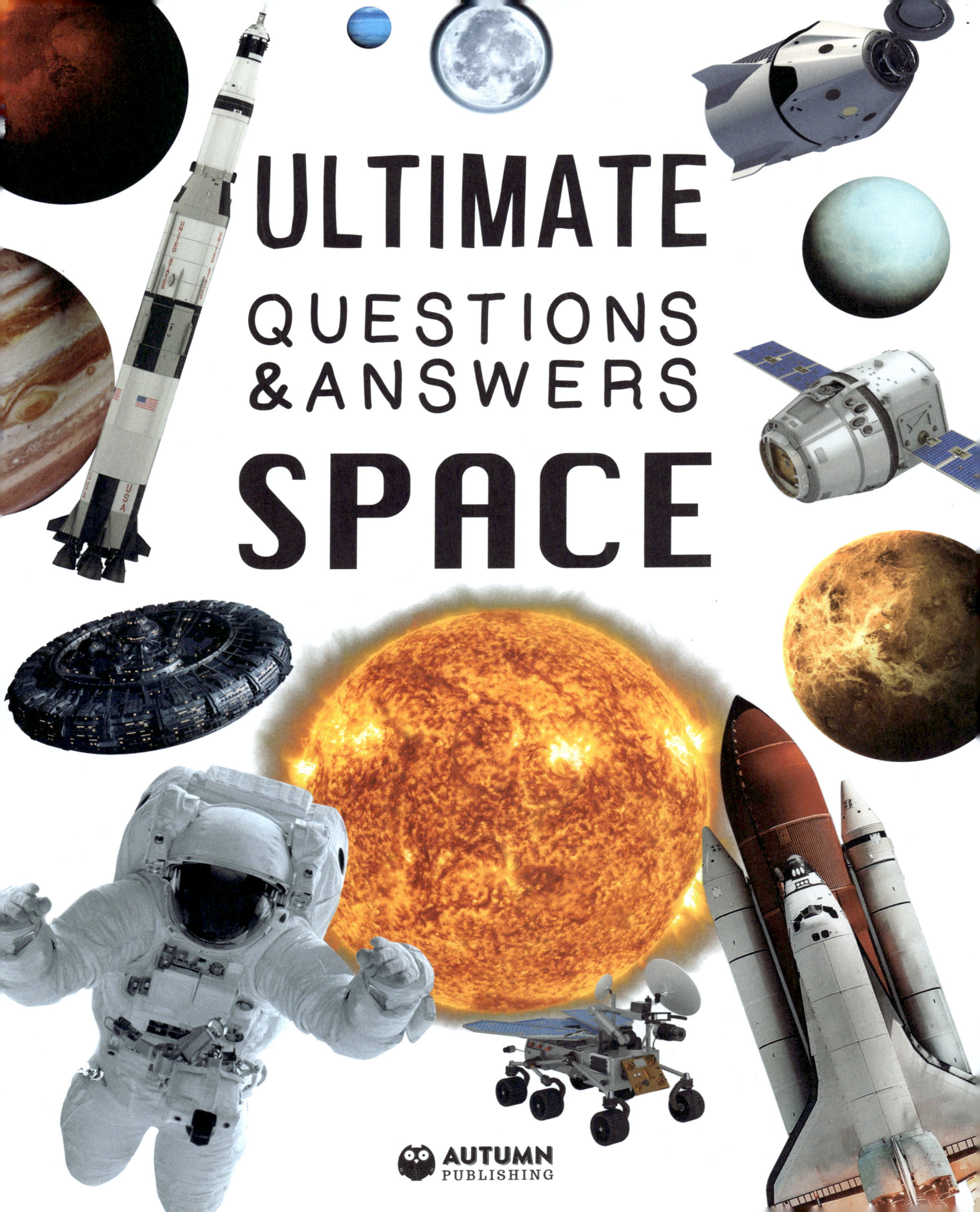

CONTENTS

OUR UNIVERSE...4

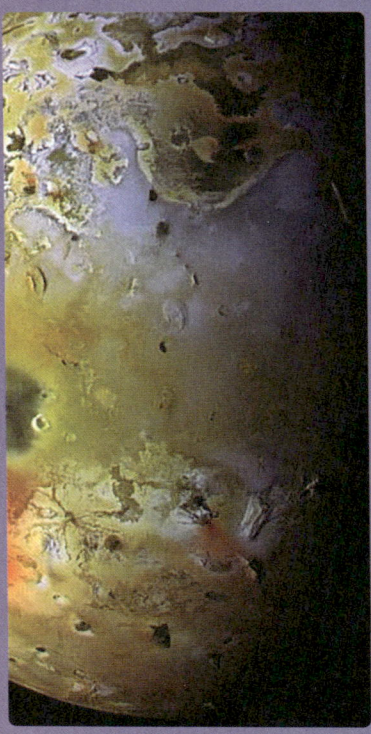

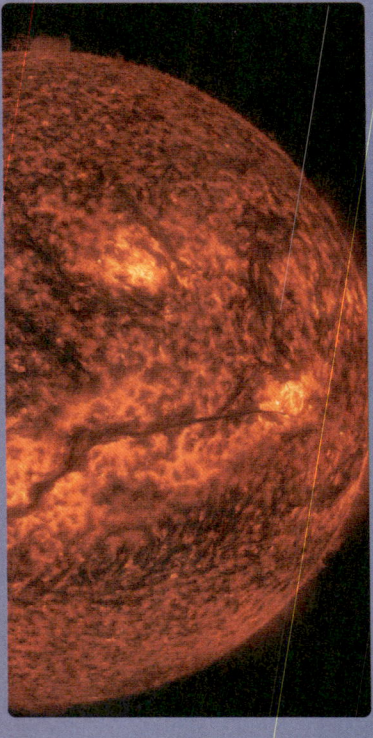

STARS...18

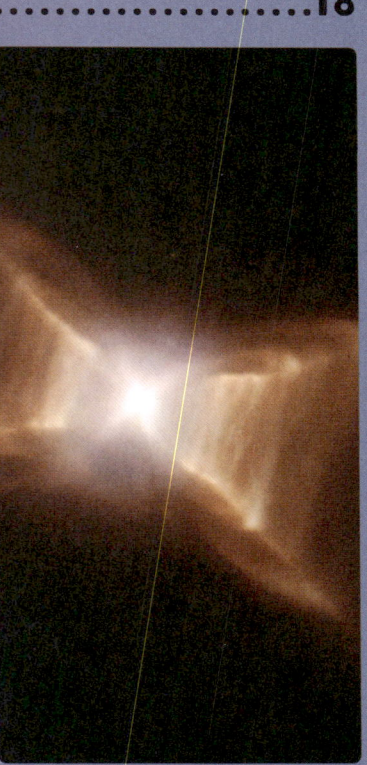

SOLAR SYSTEM...32

OUR PLANET..56

BIRTH OF THE UNIVERSE

WHAT IS THE UNIVERSE?

Scientists use the word 'Universe' to describe absolutely everything that physically exists. They believe that the Universe was created by the Big Bang. The early Universe was very small, but it contained all the matter and energy that exist in the Universe today – a dense, chaotic mix of tiny particles and forces.

Stars, nebulae and galaxies are all part of the Universe.

Big? WHAT WAS THE BIG BANG?

At the start of the Universe, probably about 13 billion years ago, there was a small, unimaginably hot and dense ball, out of which the entire Universe burst into existence with the greatest explosion of all time – the Big Bang! The Universe rapidly expanded, allowing first energy and matter, then atoms, gas clouds and galaxies to form. This continues today, so the Universe is becoming bigger and bigger.

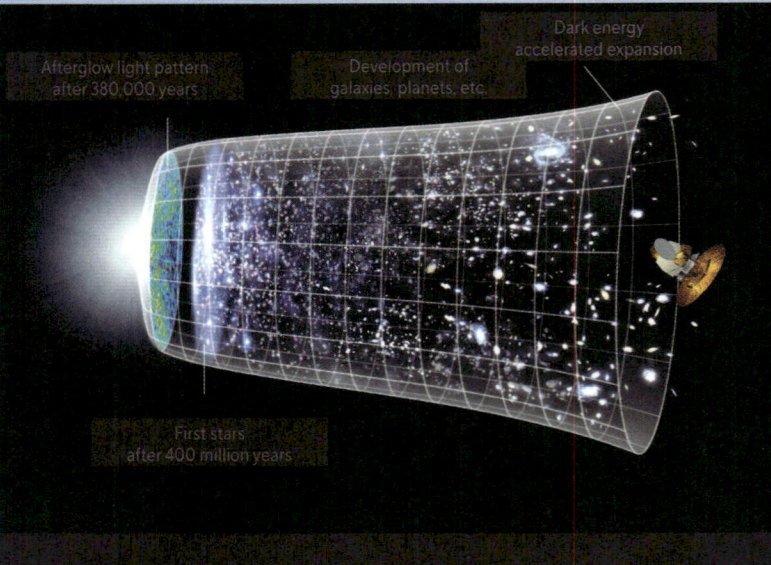

The Big Bang expansion

Afterglow light pattern after 380,000 years

Development of galaxies, planets, etc.

Dark energy accelerated expansion

First stars after 400 million years

OUR UNIVERSE | 5

Stars, planets and gas clouds

WHAT IS THE UNIVERSE MADE FROM?

The stars and gas clouds in space are made almost entirely of hydrogen and helium. Rocky planets, such as Earth, are formed from densely packed elements including carbon, oxygen, silicon, nitrogen and iron.

WHAT SHAPE IS THE UNIVERSE?

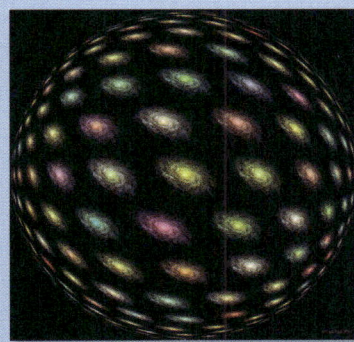
Could the Universe look like this?

Scientists are not yet sure. Just as there was once a time when the Earth's shape was debated, there is still no agreement on whether the Universe is flat, curved or spherical.

Quick-FIRE?

WHO WAS THE FIRST PERSON TO SUGGEST THE BIG BANG THEORY?

A Belgian priest, named Georges Lemaître, in 1920.

Georges Lemaitre

CAN YOU SEE THE BIG BANG?

No, but astronomers can see the afterglow of the Big Bang as low-level microwave radiation called 'microwave background radiation'. It is moving towards us from all over space.

HOW HOT WAS THE BIG BANG?

As the Universe grew from smaller than an atom to the size of a football, it cooled from infinity to 10 000 000 000 000 000 000 000 000 000 °C.

HOW LONG DID THE ORIGINAL UNIVERSE LAST?

The original Universe lasted only for a split second – just three-trillionths of a trillionth of a trillionth of a second!

HOW DO WE KNOW THE UNIVERSE IS EXPANDING?

We can work this out because distant galaxies are zooming away from us. Yet, it is not that the galaxies themselves are moving – it is the space between them that is stretching.

WHAT IS AN ORBIT?

It is a path that an object in space takes around another object. Earth orbits the Sun, along with other planets.

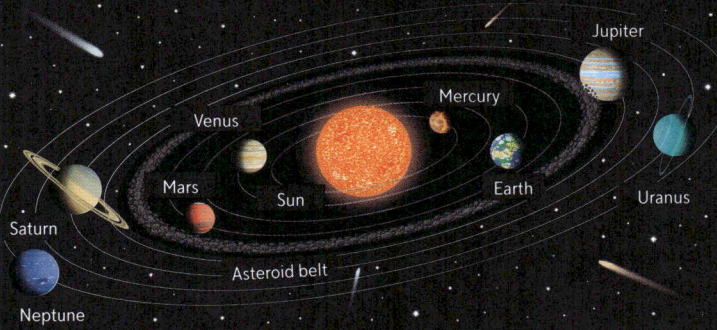
Eight planets move around the Sun.

WHAT WAS THERE BEFORE THE UNIVERSE?

No one really knows! Some people think there was a vast sea, beyond space and time, full of potential universes continually bursting into life. Ours was one of the successful ones. There may have been others and they may still exist.

Age of the Universe

WHAT IS THE BIG CRUNCH?

Some scientists think that if there is a large amount of dark matter in the Universe, it may cause it to shrink and collapse. They call this the Big Crunch.

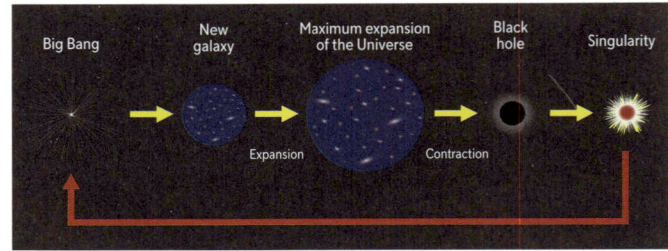

The Big Crunch

HOW DO SCIENTISTS KNOW WHAT THE EARLY UNIVERSE WAS LIKE?

No one knows for certain, but scientists have worked out possibilities through mathematical calculations based on the laws of physics. Huge, sophisticated machines called colliders and particle accelerators are used in experiments. These recreate conditions that may have existed in the early Universe by using magnets to propel particles to move at unbelievable speeds in a special tunnel, and then smashing them all together.

OUR UNIVERSE | 7

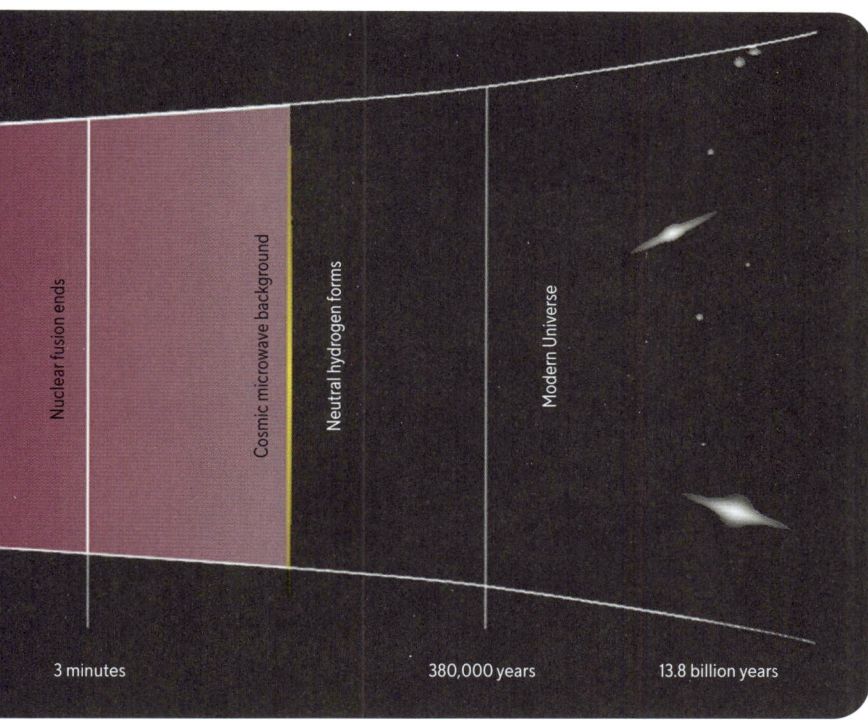

Quick-FIRE?

WHAT WAS INFLATION?

It was the dramatic expansion and cooling that took place just a fraction of a second after the Big Bang.

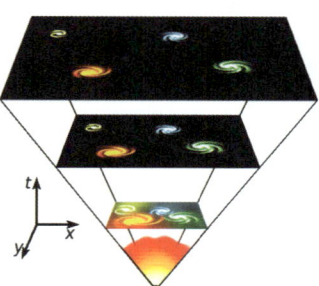

Expansion of the Universe

WHAT IS THE BIGGEST THING IN THE UNIVERSE?

The Hercules-Corona Borealis Great Wall – a sheet of galaxies that measures around 10 billion light years across.

HOW OLD IS THE UNIVERSE?

About 13 billion years old.

HOW BIG IS THE UNIVERSE?

Bigger than you can possibly imagine! The below image shows the galaxy named GN-z11, which is 32 billion light years away. It's one of the farthest groups of objects astronomers have seen.

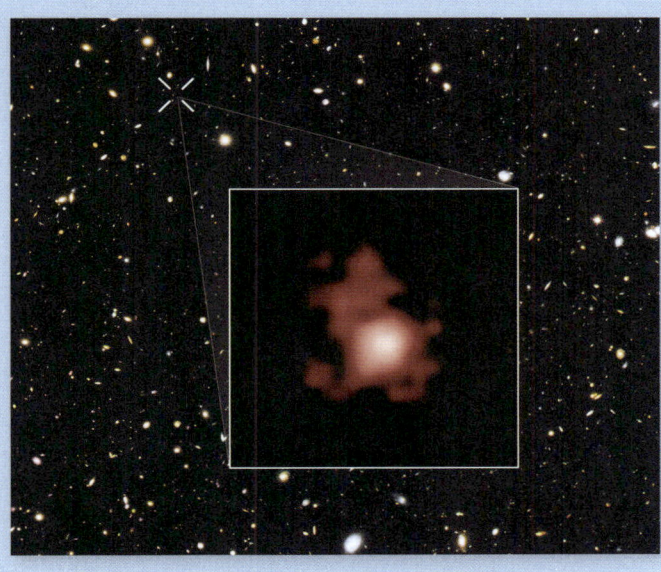

The distant galaxy GN-z11

WHAT IS THE FUTURE OF THE UNIVERSE?

Nobody knows if the Universe will keep on growing. It all depends on how much matter it contains. It is possible that gravity will stop its expansion, and there could be a contraction again – ending up in a Big Crunch! On the other hand, it may be a flat Universe, which just about avoids contraction or collapse, or an open Universe, which is forever expanding.

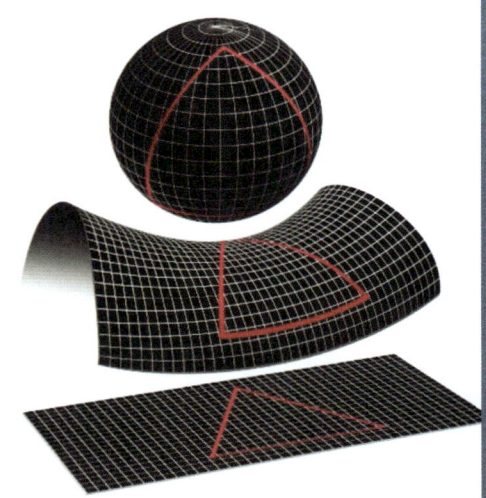

A flat Universe would be like this

WHAT IS MATTER?

Every substance in the Universe, from the tiniest speck of dust to the largest star, is matter. Anything that has mass and takes up space, is matter. All matter is made up of tiny particles. These particles are the building blocks of the Universe. There are three different forms, or states, of matter: solid, like a brick; liquid, like water; and gas, like air. Every substance can change from one state to another and back again, depending on the temperature and pressure.

Everything that exists is matter.

Quick-FIRE?

HOW CAN WE SEE QUARKS?
The paths of subatomic particles, such as quarks, can be seen when atoms collide at great speed.

When atoms collide

WHAT IS ANTIMATTER?
Antimatter is the mirror image of matter. If matter and antimatter meet, they destroy each other. Fortunately, there is no antimatter on Earth.

WHAT IS AN ATOM SMASHER?
Also known as a particle accelerator, it is a machine used to propel particles at immense speeds to study their nature.

OUR UNIVERSE | 9

HOW WERE ATOMS MADE?

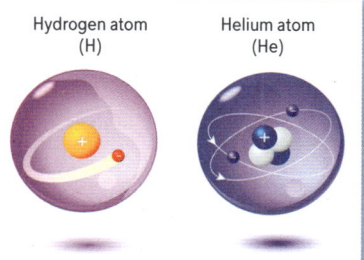
Atoms of hydrogen and helium

Atoms of hydrogen and helium were created in the early days of the Universe when quarks joined together. All other atoms were made when atoms fused together because of the intense heat and pressure inside stars.

An atom

Electron
Nucleus
Neutron
Proton

Big?
WHAT IS AN ATOM?

Atoms are the tiny bits or 'particles' of which every substance is made. They are far too small to see. Two billion atoms would fit on the dot of this 'i'! Scientists once thought they were the smallest things in the Universe, and that they were small, hard balls that could never be split or destroyed. Now they know atoms are more like clouds of energy, and are mostly empty space, dotted with even tinier subatomic particles. Right in the centre of every atom, like a pea in a football, is a dense core, or nucleus, containing two kinds of tiny particles: protons and neutrons. Around the nucleus are even tinier particles called electrons, whizzing around at the speed of light.

WHAT ARE PARTICLES?

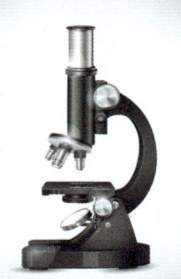
A microscope

Particles are the basic units of matter that make up everyday objects. There are hundreds of kinds of particles, but all, apart from atoms and molecules, are too small to see, even with the most powerful microscope.

WHAT IS A QUARK?

Structure of a quark

Quarks are tiny particles, much smaller than atoms. They were among the first particles to form at the birth of the Universe.

HOW SMALL IS A QUARK?

The smallest particle inside the nucleus is a quark. It is less than 10^{-20} m across. A line of 10 million billion of them would be less than a metre long.

Quarks in an atom's nucleus

WHAT IS GRAVITY?

Gravity is the mutual attraction between every bit of matter in the Universe. The more matter there is, and the closer it is, the stronger the attraction. A big planet pulls much more than a small one, or a planet that is far away. The Sun is so big that it makes its pull felt over millions of kilometres.

Gravity holds the Universe together.

The effect of Earth's gravity

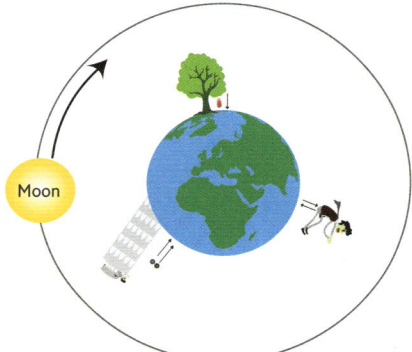

Big? WHY IS GRAVITY IMPORTANT?

Gravity is the force that holds the Universe together. It is the force that keeps Earth in one piece, keeps us on the ground, keeps planets circling the Sun, and draws the stars together in galaxies. Yet, gravity can also be so powerfully destructive that it can squeeze stars into nothing and suck galaxies into oblivion.

HOW DOES EARTH STAY IN ORBIT?

Earth stays on the same path around the Sun because it is travelling through space too fast to be drawn in by the pull of the Sun. Yet, Earth is not fast enough to overcome the Sun's pull altogether and to hurtle off into space.

Gravity holds Earth in orbit.

OUR UNIVERSE | 11

HOW DOES GRAVITY AFFECT A DYING STAR?

When some giant stars burn out, there is nothing to hold them up and they start to collapse under the force of their own gravity. As they shrink, they get denser and denser, and their gravity becomes even stronger. Ultimately the tremendous force of gravity reduces them to a size of a few kilometres.

DOES GRAVITY CAUSE DESTRUCTION IN OUR SOLAR SYSTEM?

Yes, this happens on one of Jupiter's moons, called Io. As Io revolves, the giant planet's forceful gravity pulls and sucks so powerfully that the insides of the moon heat up to the extent that volcanoes burst onto the moon's surface.

Jupiter's moon, Io

Quick-FIRE ?

Sir Isaac Newton

WHO DISCOVERED GRAVITY?

Sir Isaac Newton, in the 17th century, when he wondered why an apple fell straight to the ground and did not float away.

Launching a rocket

WHY DOES ONE NEED A ROCKET TO GO INTO SPACE?

Only a rocket can provide the power needed to escape the pull of gravity.

DO ALL PLACES ON EARTH HAVE THE SAME GRAVITATIONAL PULL?

No, because the mass of Earth is not evenly distributed in all places.

WHAT ARE THE GRACE MISSIONS?

The Gravity Recovery and Climate Experiments measures the change in distance between two identical satellites flying some distance apart to find out variations in Earth's gravity.

WHAT HAPPENS IF A SATELLITE IS LAUNCHED TOO SLOW OR TOO FAST?

If launched too slowly, the satellite falls back to Earth. If launched too fast, the satellite zooms off into space. When launched at the right speed and at exactly the right angle, the satellite is placed successfully in orbit.

Spacecraft in orbit

GALAXIES

WHAT IS A GALAXY?

A galaxy is a cluster of billions of stars. Galaxies formed from clouds of hydrogen and helium gas as concentrations of the gases within the clouds drew together. The youngest-known galaxy is just 500 million years old. There are three main types of galaxy: spiral, elliptical and irregular.

The Milky Way galaxy

WHAT ARE ELLIPTICAL GALAXIES?

Galaxies that are shaped like rugby balls. There is no gas or dust remaining in an elliptical galaxy, so no new stars can form.

The elliptical galaxy Centaurus

Big? WHAT ARE SPIRAL GALAXIES?

Spiral galaxies look like spinning Catherine wheels. They have a dense core of stars, surrounded by long whirling arms. Our Milky Way is one of them. It has between 100 and 400 billion stars, arranged in a shape like a fried egg, and spread over 100,000 light years. One of these stars is the Sun. This huge 'star city' is called the Milky Way Galaxy because we see it as a pale, white band across the night sky.

The Milky Way seen from Earth

OUR UNIVERSE | 13

HOW BIG CAN GALAXIES BE?

Galaxies are absolutely gigantic. Although they contain millions, or even trillions, of stars, they are mostly empty space, as the stars are far apart. If each star was a person, the nearest neighbour would be almost as far away as the Moon!

Much of space is empty.

Quick-FIRE?

WHAT ARE BARRED SPIRAL GALAXIES?

Spiral galaxies, with a central bar from which arms of stars trail like water from a spinning garden sprinkler.

Barred galaxy NGC 986

HOW MANY GALAXIES ARE VISIBLE TO THE NAKED EYE?

Three galaxies, apart from the Milky Way: the Large and Small Magellanic Clouds, and Andromeda.

Gas clouds and stars in the Large Magellanic Cloud

HOW MANY GALAXIES ARE THERE?

At least 200 billion, possibly up to two trillion!

WHY ARE ONLY A FEW GALAXIES VISIBLE TO US?

The Universe is believed to be ever-expanding. From Earth, astronomers observe that galaxies are moving away all the time and becoming more distant. Though galaxies contain many millions of stars, they are so far away that even with a reasonably good telescope, most look like blurs in the night sky.

WHAT ARE IRREGULAR GALAXIES?

Galaxies that have no particular shape at all. Most irregular galaxies were once spiral or elliptical, but have been pulled apart by gravity.

The irregular galaxy NGC 4485

NEBULAE

Orion Nebula

WHAT ARE NEBULAE?

Stars start life in giant clouds of dust and gas called nebulae. They are born in these clouds as gravity pulls the dust into clumps so intensely that the pressure makes them glow. On a clear night, fuzzy patches of light can be seen among the stars. Some of these patches are distant galaxies, but others are nebulae, many times bigger than any star.

WHAT IS THE COALSACK NEBULA?

It is a dark nebula, about 500 light years from Earth, in the constellation of the Southern Cross. Coalsack can be identified with the naked eye as it blots out a section of stars in the Milky Way. The Australian Aboriginal people observed it and likened it to the head of an emu in the sky.

WHAT ARE THE DIFFERENT TYPES OF NEBULAE?

The dark nebula Coalsack

All nebulae are forms of matter found between stars (interstellar). Their appearance varies according to their temperature and density, and where they are located in relation to the observer. On the basis of how they look, nebulae are divided into two broad types: dark nebulae and bright nebulae. Dark nebulae appear as irregularly shaped dense black patches in the sky, which blot out the light of the stars that lie beyond them. Bright nebulae appear as faintly glowing surfaces.

WHICH IS THE **NEBULA CLOSEST TO EARTH?**

The Helix Nebula, 650 light years away, in the constellation of Aquarius. Helix is a planetary nebula, which is actually the remains of a dying star, like our Sun. It has taken three powerful space telescopes to map Helix's glowing structure: GALEX, Spitzer and the Wide-Field Infrared Survey Explorer (WISE).

Helix Nebula

Quick-FIRE?

WHERE ARE THE PILLARS OF CREATION?

In the Eagle Nebula.

The Pillars of Creation

CAN NEBULAE BE RECTANGLE-SHAPED?

Yes, the Red Rectangle Nebula is an example.

Red Rectangle Nebula

WHICH NEBULA HAS A GREENISH TINT?

The Orion Nebula, one of the celestial bodies that has been best studied by astronomers.

WHY ARE NEBULAE CALLED **STAR NURSERIES?**

The star-forming region of the Eagle Nebula

The gases and dust that make up nebulae are drawn closer together in clumps by the pull of gravity. These clumps become dense and develop even denser cores that are very hot. These cores are the beginning of new stars, which is why nebulae are called star nurseries.

WHAT ARE **DIFFUSE NEBULAE?**

Nebulae are of various types. Diffuse nebulae are those that do not have well-defined boundaries. One example is the Eta Carinae, 7,500 light years away from Earth. Two kinds of diffuse nebulae are emission nebulae, which give out their own light; and reflection nebulae, which are visible because they reflect the light of other stars.

Eta Carinae

WHAT IS A BLACK HOLE?

A black hole is a region in the Universe that has such an immense gravitational pull that it acts like a funnel and sucks in space. Not even light can escape the pull of a black hole, which is why it is called black. The hole's interior cannot be seen.

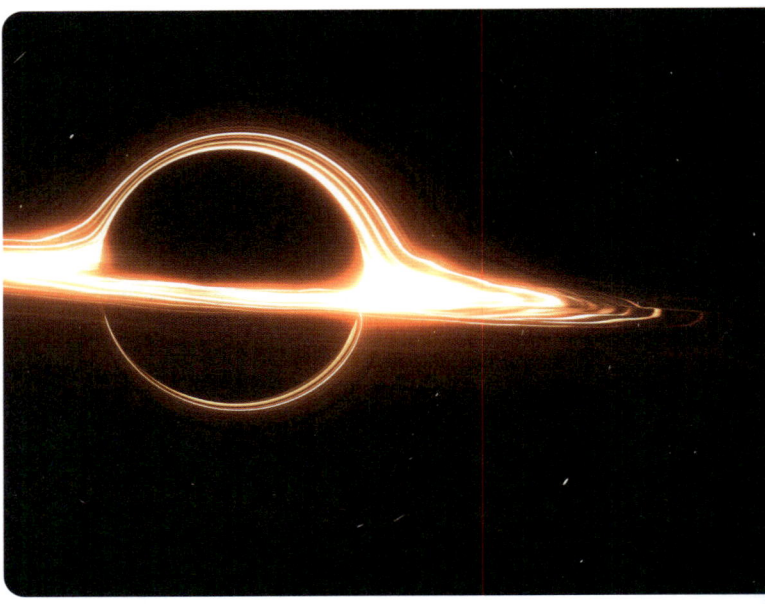

Artist's view of a black hole

WHAT HAPPENS INSIDE A BLACK HOLE?

Nothing that goes into a black hole comes out. Gravity inside it is so intense that it bends space itself. Space and time, as we know them, do not exist in a black hole. The laws of physics do not apply here.

HOW CAN WE SEE A BLACK HOLE?

The black hole contains so much matter in such a small space that its gravitational pull even drags in light. Scientists can make out the presence of a black hole from the disc of light around it, which is the radiation emitted by stars being ripped to shreds as they are sucked in. A giant black hole called Sagittarius A* exists at the centre of our galaxy.

Quick-FIRE?

HAS A BLACK HOLE EVER BEEN SEEN?

The first image of a black hole, lurking at the centre of the Messier 87 galaxy, was created in 2019 by the Event Horizon Telescope observatories.

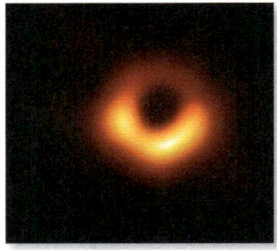

Black hole, Messier 87 galaxy

WHERE ARE BLACK HOLES FOUND?

Supermassive black holes are at the centre of galaxies, but smaller black holes could be anywhere in a galaxy.

HOW MANY BLACK HOLES ARE THERE?

No one really knows, but scientists think there may be as many as 10 million to one billion black holes in the Milky Way alone.

Sagittarius A*

OUR UNIVERSE | 17

ARE QUASARS ALSO IN BLACK HOLES?

The brightest objects in the Universe, called quasars, could be compared to the screams of decaying stars – the intense radiation from matter being ripped to shreds as it is sucked into a black hole.

A quasar

HOW BIG IS A BLACK HOLE?

The size of the black hole is usually taken to be the size of the volume of space from which light cannot escape. The black hole at the heart of our galaxy is about 23.6 million km wide, and has the mass of 4 million Suns! But there could be many black holes that are much bigger.

Black hole over a star field

HOW IS A BLACK HOLE FORMED?

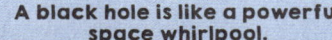

A black hole is like a powerful space whirlpool.

When a large star explodes as a supernova, the centre of the star is violently compressed by the shock of the explosion. As it compresses, it becomes more dense and its gravity becomes increasingly powerful. Ultimately, the gravity is sucked into an impossibly small point called a singularity. Gravity around the singularity is so ferociously powerful that it not only sucks in light, but even bends time and space.

WHAT ARE STARS?

Stars are huge fiery balls of gas. They shine because they are burning. Deep inside, as they are squeezed by the star's gravity, hydrogen atoms fuse together to form helium. This nuclear reaction unleashes so much energy that temperatures at the core of the star reach millions of degrees, making the surface glow and sending out light, heat, radio waves and many other kinds of radiation.

A bright, burning star

Quick-FIRE?

HOW MANY STARS ARE BORN IN THE UNIVERSE IN A YEAR?

About 150 billion. That means 4,800 stars are being born every second!

A star-crowded sky

Sirius A

WHEN DID THE FIRST STARS FORM?

It was when the Universe was only around 200 million years old.

HOW FAR IS THE BRIGHTEST-LOOKING STAR IN THE SKY?

Sirius A, the brightest-looking star, is 8.6 light years away from Earth.

DO STARS HAVE COLOUR?

They do. A star could be red, orange, yellow, blue or white, depending on how hot it is. The almost white-looking Sun is a little cooler than blue stars, which are hottest. Red stars are the coolest.

Stars in the Milky Way

STARS

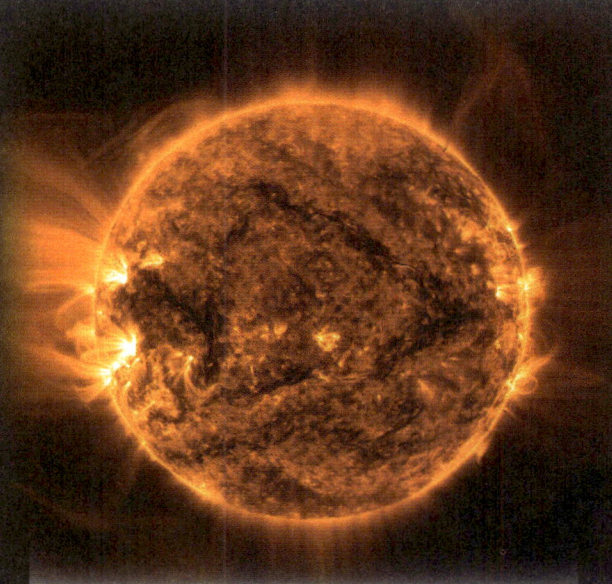

The Sun

HOW DOES A STAR BURN STEADILY?

In medium-sized stars, like our Sun, the heat generated in the core pushes gas out as hard as gravity pulls it in, so the star burns steadily for billions of years. But stars are being born and dying all the time, all over the Universe. The bigger the star, the shorter its life.

The Monkey Head Nebula produces many stars

Big? HOW AND WHERE ARE STARS BORN?

Stars are born in the biggest nebulae, the vast clouds of dust and gas stretched throughout space. These clouds are 99 per cent hydrogen and helium, with tiny amounts of other gases, and minute quantities of icy, cosmic dust.

Stars are born when clumps of gas in space are drawn together by their own gravity and the middle of the clump is squeezed so hard that temperatures reach 10 million °C. This starts a nuclear fusion reaction and the heat makes the star shine.

HOW CLOSE ARE WE TO THE STARS?

The star nearest to us (apart from the Sun), the faint Proxima Centauri, is part of the triple-star system called Alpha Centauri, which is visible to the naked eye as a single star. The stars are all so distant that we can see them only as pinpoints of light in the night sky.

Proxima Centauri

HOW ARE STARS CLASSIFIED?

The way that stars are sorted into groups depends on their size and how brightly they burn. Large stars burn their fuel resources very quickly and are short-lived. Small stars burn their fuel slowly and can last for billions of years. Most stars are a similar size to our Sun. But some are giants, 100 times bigger, or supergiants, over 300 times larger! Dwarf stars are very small – smaller than the Earth.

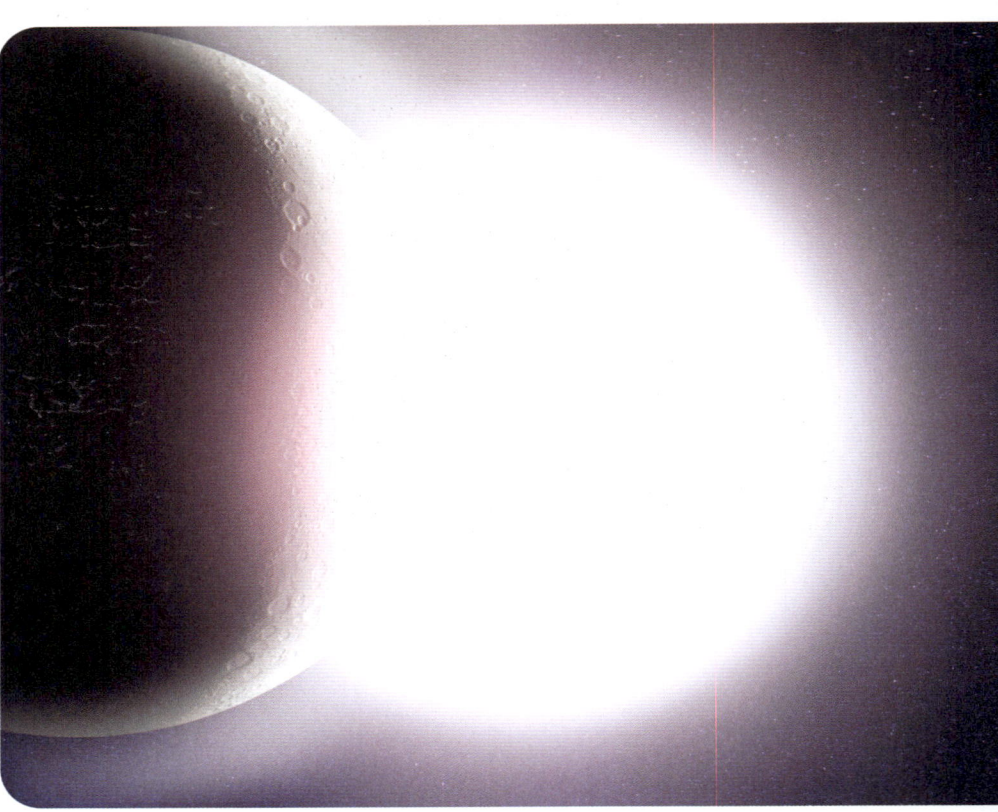

Cygnus OB2-12 with a planet

Big? WHAT ARE RED GIANT STARS?

HOW SMALL ARE DWARF STARS?

Dwarf stars are actually very small – smaller than Earth. Very old stars can shrink under the power of their own gravity to even tinier neutron stars – no bigger than the city of London, but can be so dense that they are as heavy as the Sun!

The dwarf star TRAPPIST-1 and the Sun

A red giant

Red giants are 20 to 100 times as big as the Sun. A red giant is a huge, cool star. It forms as the surface gas on a medium-sized star, nearing the end of its life, swells up. There are also supergiants, like Betelgeuse, over 500 times as big and 700 million km across. Despite being much brighter than our Sun, the hypergiant Cygnus OB2-12, appears quite dim, as it is over 5,000 light years away!

Quick-FIRE ?

WHAT IS THE **HABITABLE ZONE OF A STAR?**
The distance a planet should be from the star it orbits so that it has the right temperature to have liquid water and, therefore, the possibility of life.

WHEN **DOES A STAR ABSORB ENERGY** INSTEAD OF RELEASING IT?
When iron forms in its core.

Double stars

WHAT ARE **DOUBLE STARS?**
Our Sun is alone in space. But many stars have one or more nearby companions. Double stars are called binaries.

WHICH IS THE BIGGEST STAR?

It is difficult to say as there are so many stars in the Universe and new stars keep being discovered. Until some time ago, VY Canis Majoris, 5,000 light years away, was possibly the biggest known, at over 1,500 times the size of the Sun. UY Scuti, in the constellation of Scutum, is over 1,700 times the size of the Sun, while NML Cygni, another red hyper-giant could be way over 2,000 times the Sun's size!

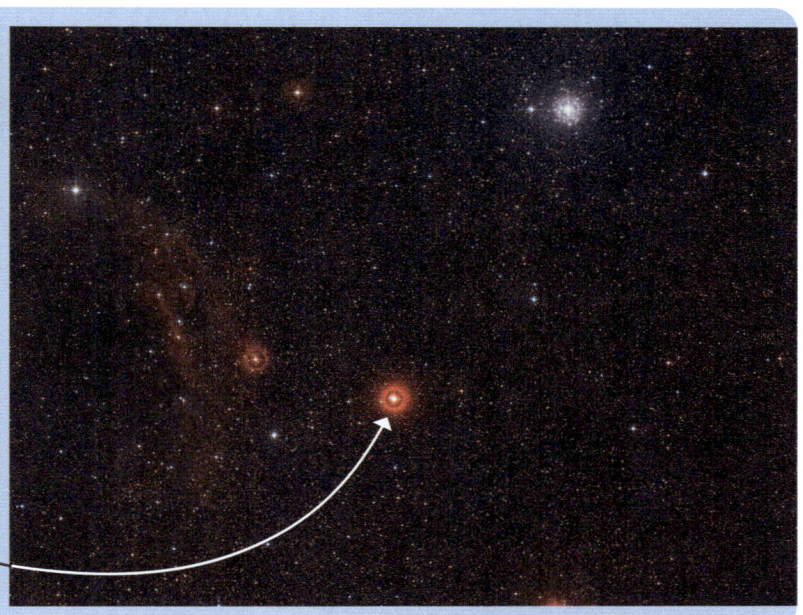

The giant red star VY Canis Majoris

WHICH **STARS THROB?**

The light from variable stars flares up and down. Cepheid stars are big, old stars that pulsate over a few days, or a few weeks. RR Lyrae are old, yellow stars that pulsate over a few hours.

Cepheid variable star RS Puppis

HOW FAR TO THE NEAREST STAR?

The Universe is very, very big. The distances between stars are so huge that astronomers don't measure them in kilometres, but in light years – which is how far light, the fastest thing in the Universe, with a speed of about 300,000 km per second, travels in a year. A light year is 9,460 billion km, and the nearest star is about 4.3 light years away!

Stars in the night sky

HOW ARE DISTANCES IN SPACE CALCULATED?

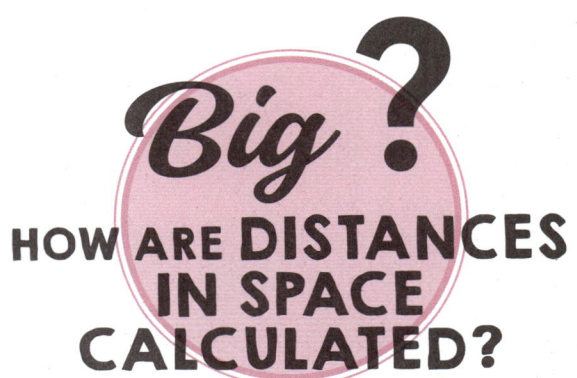

Using parallaxes

Astronomers may measure large distances in parsecs. A parsec is 3.26 light years. Parsec distances are worked out by using the geometry of parallaxes – the way a star seems to shift slightly in position in the night sky as Earth moves around the Sun. Distances to the nearer stars can be worked out also by estimating their brightness. Astronomers can estimate from the colour of a star how bright it should look compared to other stars. Radar beams are used for nearby objects, like the Sun and the planets.

WHAT IS THE FARTHEST OBJECT WE CAN SEE WITH THE NAKED EYE?

The Andromeda Galaxy, about 2.5 million light years away. It is the most massive in the local group of galaxies, which includes our Milky Way. Andromeda is visible as a smudge in the night sky.

Andromeda Galaxy

STARS | 23

Quick-FIRE ?

Which IS THE NEAREST STAR, OTHER THAN THE SUN?

Proxima Centauri, about 4.24 light years, or 40 trillion km, away.

Proxima Centauri

HOW FAR DO THE ANDROMEDA GALAXY'S STARS SPREAD OUT?

About 220,000 light years across.

Andromeda's stars

HOW SOON COULD YOU GET TO THE SUN AND BACK AT THE SPEED OF LIGHT?

In 16 minutes.

WHAT ARE QUASARS?

A class of objects called active galactic nuclei, quasars are the farthest objects we can see with telescopes – the farthest has been detected at 13 billion light years away. Quasars probably surround black holes, and their enormous energy makes them many times brighter than all other nearby stars.

WHAT IS REDSHIFT?

When a galaxy is moving away from us, the waves of light coming from it become stretched out and they look redder. The greater this 'redshift', the faster the galaxy is moving. When a galaxy is moving towards us, the waves are compressed and they appear blue.

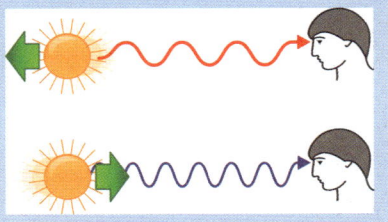

Redshift vs blueshift

WHAT ARE STANDARD CANDLES?

When measuring the distance to middle-distance stars, astronomers compare the star's brightness to the brightness of stars that they know, or 'standard candles'. The dimmer the star looks in comparison, the farther away it is.

WHAT ARE CONSTELLATIONS?

Constellations are small patterns of stars in the sky. Each pattern has its own name. They are often named after the object or figure that they resemble, such as a lion or a cross. The stars in a constellation may not be very close to each other in reality – they only appear to be close when viewed from Earth.

Centaurus constellation

Looking at Orion in space

WHAT IS THE HUNTER?

The constellation of Orion is named after a hunter of the same name in Greek mythology. The constellation looks like a hunter holding a sword. The hunter's head, shoulders, three-starred belt, legs and sword can be seen. Orion is one of the brightest constellations in the night sky.

Quick-FIRE?

WHAT IS THE LARGEST CONSTELLATION?
Hydra. It is named after a many-headed water snake.

Hydra

SINCE WHEN HAVE PEOPLE GROUPED STARS INTO CONSTELLATIONS?
At least 10,000 years ago. Some cave paintings in Europe from that time show groups of stars like the ones we see today.

HOW DID CONSTELLATIONS HELP TRAVELLERS?
Sailors observed them to plot their way across the seas.

HOW MANY CONSTELLATIONS ARE THERE?
Eighty-eight.

How are STARS IN A CONSTELLATION LINKED?

There is no real link between stars in a constellation - the patterns are totally imaginary. But without constellations to help us, it would be hard for us to locate stars in the sky since there are as many as 1,500 stars visible on a dark, clear night. If you can recognise the constellation Orion by its pattern, you can immediately identify the stars in Orion's belt, shoulder and foot. You can even use Orion as a reference to identify other constellations in its neighbourhood.

Orion

WHERE IS THE POLE STAR?

The Pole Star is a bright star that lies directly over the North Pole. It is also called Polaris or Alpha Ursae Minoris, as it is the brightest star in the Ursa Minor constellation. When photographed with a long exposure, all the stars seem to rotate about the Pole Star as the Earth turns.

Polaris, the Pole Star

WHAT ARE ALPHA AND BETA?

The stars in each constellation are named after letters in the Greek alphabet. Usually, the brightest star in each constellation is called Alpha, the next brightest Beta, and so on. So Alpha Centauri is the brightest star in the constellation Centaurus.

WHAT IS THE PLOUGH?

Also known as the Big Dipper, the Plough is the most visible part of the Great Bear or Ursa Major constellation, which is one of the best-known constellations in the northern sky. Drawing an imaginary line between Dubhe and Merak, two stars of the Plough, we can locate the Pole Star.

WHAT ARE STAR GROUPS OR CLUSTERS?

Stars are rarely entirely alone within a galaxy. They usually exist in multi-star groups, or clusters, which are held together by the force of each other's gravity. Globular clusters are big and round, and contain hundreds of thousands, sometimes millions, of ancient stars. Open or galactic clusters are formless and small, made up of a few hundred or thousand stars.

A cluster of ancient stars

Big? DOES THE MILKY WAY HAVE STAR CLUSTERS?

The Milky Way

There are more than 150 known globular star clusters in the Milky Way. About one-third of them are located near the centre of the galaxy. Some clusters have masses equal to one million Suns. Some are as old as the Universe itself. Most of the open clusters lie in the spiral arms of the galaxy and can be used to discern the shape of the galaxy. The brightest stars in the open clusters are 150,000 times brighter than the Sun!

CAN TWO STARS EXIST TOGETHER?

Our Sun is alone in space, but many stars have one or more nearby companions. These are called binaries. In fact, at least 80 per cent of the stars we see in the sky as single pricks of lights are actually binaries. The brighter of the two stars in a binary system is called 'primary', while the dimmer one is known as 'secondary'.

Artist's view of the binary star system AR Scorpii

WHAT ARE THE PLEIADES?

The Pleiades are over 1,000 stars grouped together – seven of which are visible to the naked eye – that formed in the same cloud of dust and gas. The stars are held loosely together by gravity. The constellation Pleiades is named after the seven daughters of Atlas in Greek mythology.

Pleiades

Quick-FIRE?

Star cluster

WHAT KIND OF STARS ARE IN GLOBULAR CLUSTERS?
Very old stars and small stars.

At the edge of the galaxy

WHAT DO STAR CLUSTERS TELL US ABOUT OUR SOLAR SYSTEM?
That it isn't located near the centre of the galaxy, as was earlier thought, but nearer to its edge.

WHAT HAVE WE LEARNT FROM STAR CLUSTERS ABOUT THE AGE OF THE UNIVERSE?
That it is about 13.7 billion years old.

WHAT IS THE NAME OF THE COMPANION STAR OF SIRIUS, THE DOG STAR?
The Pup!

WHAT MAKES STARS GLOW?

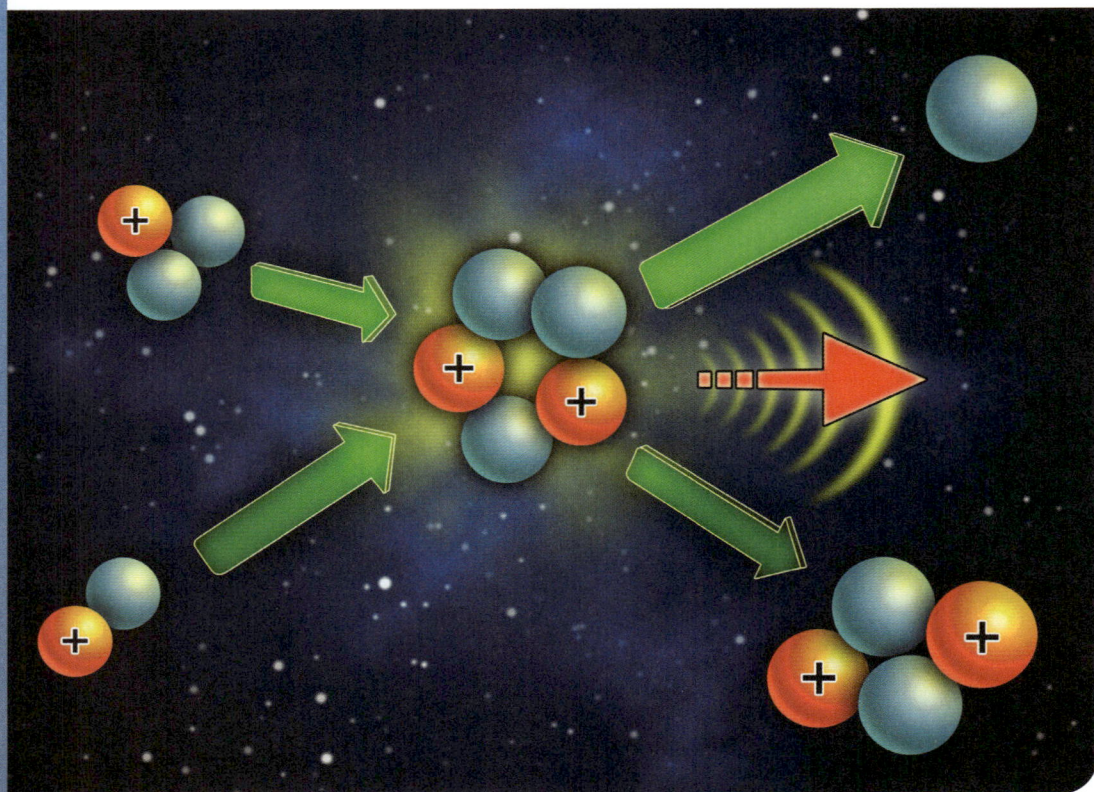

Stars glow because the enormous pressure deep inside them generates nuclear fusion reactions in which hydrogen atoms are joined together, releasing huge quantities of energy.

Stars release nuclear energy.

Quick-FIRE?

WHAT IS A RED STAR?
Older stars are often cooler and dimmer and take on a reddish glow.

Red stars are older and cooler

WHAT ARE WHITE STARS?
Young stars that often burn hot and bright and can be seen as blue-white lights in the night sky.

WHAT DOES STARLIGHT TELL US?
The colour of starlight tells us how hot a star is and also how old it is.

WHY DO STARS TWINKLE?

It is because Earth's atmosphere is never still, and starlight twinkles as the air wavers. Light from the nearby planets is not distorted as much, so they don't twinkle.

Starlight is distorted by Earth's atmosphere.

STARS 29

Wolf-Rayet stars are some of the hottest stars.

HOW **HOT** ARE STARS?

The surface temperature of the hottest, brightest stars is over 55,500 °C.

WHICH IS **THE BRIGHTEST STAR?**

Eta Carinae is a stellar system with two stars. The bigger star burns as bright as 4.7 million Suns. A close contender is the Peony Nebula star with the light of about 3.2 million Suns.

Eta Carinae

Big?
WHAT COLOUR ARE THE STARS?

The colour of a star depends on how bright it is. The Hertzsprung-Russell diagram shows the relationship between the luminosity (brightness) of a star and its temperature. The hotter stars are brighter and they glow white and blue, whereas the cooler stars are dimmer and have a redder glow. Astronomers use the Hertzsprung-Russell diagram to classify stars by their temperature, colour or luminosity.

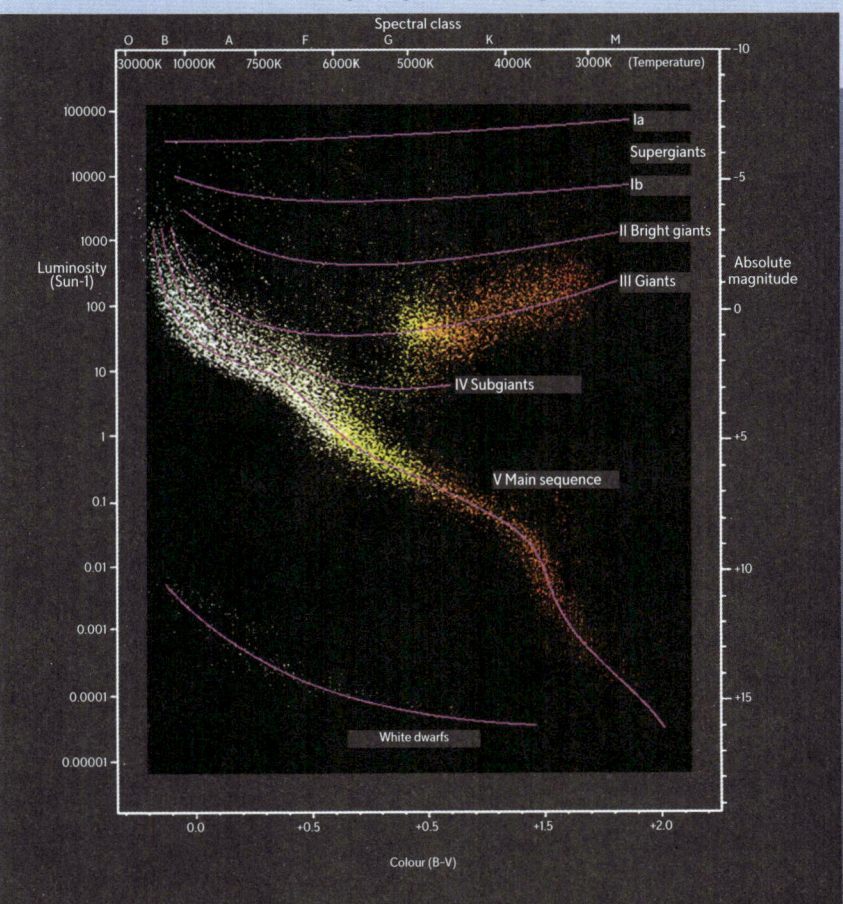

DEATH OF STARS

WHY DO STARS DIE?

Stars make energy by turning hydrogen into helium. When the hydrogen is used up, the star shrinks to burn helium. Towards the end of its life, the star uses any other available nuclear fuels. Finally, when the star's supplies of energy are all gone, it dies.

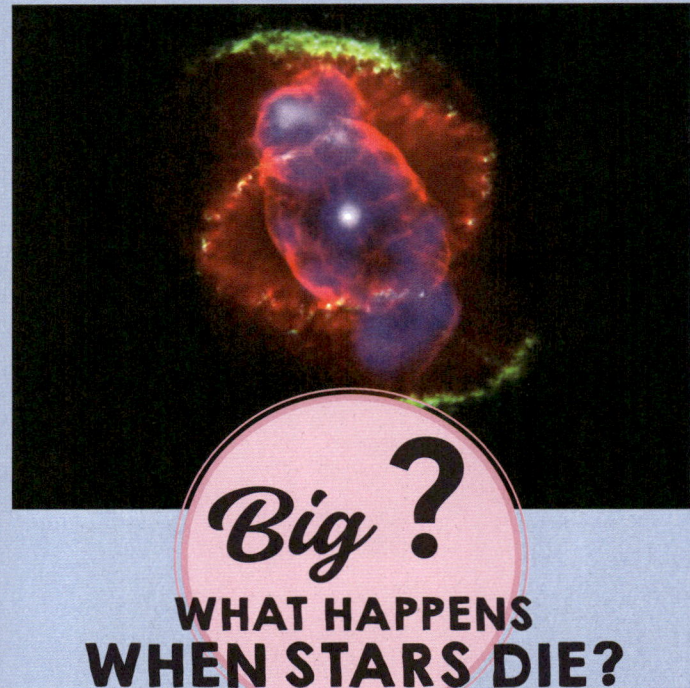

Cat's Eye Nebula is a dying star.

A gaseous halo surrounds the Cat's Eye Nebula.

Big? WHAT HAPPENS WHEN STARS DIE?

When a star has used up all its energy, it blows up, shrinks, goes cold, or becomes a black hole. When the helium runs out, the outer layers cool and the star swells to become a red giant. The biggest stars go on swelling to become supergiants, with cores so pressurised that carbon and silicon join together to make iron. Just how long it takes to reach this point depends on the size of the star.

HOW LONG DO STARS LIVE?

Stars die and new stars are born all the time. The biggest, brightest stars have lots of nuclear fuel, but live fast and die young, at the age of about 10 million years. Medium-sized stars like our Sun live for about 10 billion years. The smallest stars have little nuclear fuel, but burn slow and long. A star which is twice as big as the Sun lives only for one-tenth of the Sun's life.

STARS 31

The Sun still burns strong.

HOW WILL OUR SUN DIE?

The Sun will exhaust its supply of hydrogen fuel in about 5 billion years. Its core will collapse inwards and become hot enough to ignite its helium atoms. The Sun will then swell up to become a red giant. The outer layers will drift off, making a planetary nebula, leaving behind the core of the Sun. This will gradually cool off.

WHAT IS A SUPERNOVA?

Supernova 1987A

It is a gigantic explosion that destroys a supergiant star. For a few minutes, the supernova burns as bright as a billion suns. Supernovae are usually visible only through a telescope. But in 1987, for the first time in 400 years, a supernova (Supernova 1987A) was visible to the naked eye.

Quick-FIRE?

HOW MUCH ENERGY CAN A SUPERNOVA SEND OUT?

More in a few seconds than our Sun can in 200 million years.

A supernova burns brightly.

HOW MUCH DOES A TEASPOON OF NEUTRON STAR MATERIAL WEIGH?

About 10 billion tonnes!

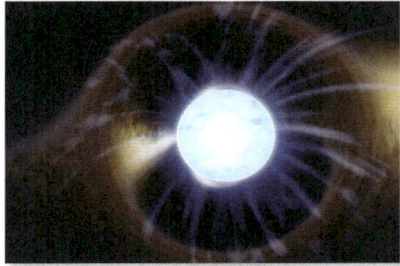

A neutron star

WHAT HAPPENS IN A SUPERNOVA EXPLOSION?

Gases burst outward at speeds between 15,000 and 40,000 km per second.

WHAT ARE LITTLE GREEN MEN?

Deep in space, there are the tiny, dense remains of supernovae, known as neutron stars. They are called pulsars or little green men (LGM). As they spin, they flash high-energy radio signals that are picked up by instruments on Earth.

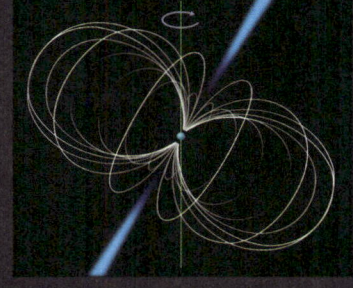

Diagram illustrating a pulsar

SOLAR SYSTEM

WHAT MAKES UP OUR SOLAR SYSTEM?

Our Solar System includes the Sun, eight planets and their moons, dwarf planets and other balls of rock, ice and metal such as asteroids and comets. The Sun is a star and makes up 99 per cent of the mass of our Solar System.

Our Solar System

Inner planets

Mercury

Mars

Earth

Venus

Big? WHAT ARE THE INNER PLANETS?

Mercury, Mars, Earth and Venus are known as the inner planets. Located inside the asteroid belt, these four planets orbit closest to the Sun. Though smaller than some of the outer planets, these four planets are made of rock and share a similar construction to an egg. Each planet has a hard 'shell', or crust of rock; a 'white', or mantle of soft, semi-molten rock; and a 'yolk', or core of hot, often molten, iron and nickel. All planets have an atmosphere of gas, which differs between each planet. These planets are all called terrestrial (Earth-like) because of their hard surface layer – this crust is what makes it possible for spacecraft to land on them.

SOLAR SYSTEM 35

WHAT ARE GAS GIANTS?

Jupiter, Saturn, Uranus and Neptune are all gas giants. This means that they are large planets entirely made up of swirling clouds of gases such as hydrogen and helium, with no rock or other hard mineral. This means that these planets do not have a solid surface area.

Jupiter, Saturn, Uranus and Neptune

Quick-FIRE?

WHO DISCOVERED URANUS?
William Herschel, in 1781.

William Herschel

WHAT IS UNIQUE ABOUT NEPTUNE'S MOON, TRITON?
It is the only moon that orbits backwards.

WHICH OF THE INNER PLANETS HAVE MOONS?
Mars and Earth.

WHICH PLANET HAS THE MOST WATER ON ITS SURFACE?
Earth.

WHAT IS THE KUIPER BELT?

It is a region of the Solar System that lies beyond Neptune. Thousands of relatively small, frozen objects orbit there.

Kuiper Belt

WHAT ARE ICE GIANTS?

Neptune and Uranus, located beyond Saturn, are the seventh and eighth planets of our Solar System. Farthest from the Sun, these gas giants are also sometimes called ice giants, because they contain a large amount of icy water, methane and ammonia.

IS OUR SUN ALSO IN ORBIT?

Our Solar System is part of a larger galaxy called the Milky Way. The Sun, which is at the centre of our Solar System, orbits around the centre of the Milky Way. It takes our Solar System about 225 to 250 million years to complete one orbit of the Milky Way.

Neptune

Uranus

WHAT IS THE SUN?

The Sun is a star – a vast, fiery spinning ball of hot gases, three-quarters hydrogen, and a quarter helium. Over a million Earths could fit inside the Sun. The gigantic mass of the Sun creates immense pressures in its core. Such huge pressures increase the temperature to 15 billion °C. All this heat turns the surface of the Sun into a raging inferno that burns so brightly that it lights Earth, over 147.5 million km away!

Nuclear fusion

WHAT MAKES THE SUN BURN?

The Sun gets its heat from nuclear fusion. Huge pressures deep inside the Sun force the nuclei (cores) of hydrogen atoms to fuse together to make helium atoms, releasing vast amounts of nuclear energy.

The blazing Sun

Big? WHAT HAPPENS INSIDE THE SUN?

Dark spots and solar flares

Pressures in the Sun's core are 2,000 billion times the pressure of Earth's atmosphere. These forces push together so many hydrogen atoms that it is as if 100 billion nuclear bombs were going off each second! The heat from the Sun's core erupts on the surface in patches called granules, and temperatures reach a super-hot 6,000 °C. Giant flame-like tongues of hot hydrogen, called solar prominences, shoot 100,000 km out into space. Every now and then, too, huge five-minute eruptions of energy called solar flares burst from the surface. Sunspots – dark blotches over 50 times the size of Africa – mark areas that are, temporarily, a little cooler.

SOLAR SYSTEM 37

HOW OLD IS THE SUN?

The Sun is a middle-aged star. It probably formed around 4.6 billion years ago. It is likely to burn for another 5 billion years, and then die in a blaze so bright that Earth will be incinerated!

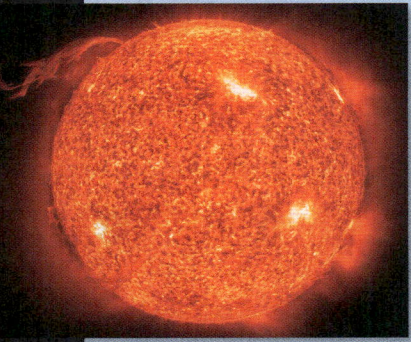
Heat from the Sun

WHAT IS SOLAR WIND?

Streaming out from the Sun at hundreds of kilometres per second are a million tonnes of electrically charged particles. This stream of particles is called the solar wind. Earth is protected from the lethal solar wind by its magnetic field.

Solar wind and Earth's magnetic field

Quick-FIRE?

WHAT IS THE SUN'S DIAMETER?

The Sun is a medium-sized star, 1,392,000 km across – that is, more than 100 times the diameter of Earth.

HOW LONG DOES A SOLAR PROMINENCE LAST?

A solar prominence forms in about a day, and may last for several months.

HOW MUCH DOES THE SUN WEIGH?

The Sun weighs 2,000 trillion trillion tonnes – over 300,000 times the weight of Earth!

WHAT IS THE SUN'S ENERGY CONSUMPTION?

The Sun burns up 4 million tonnes of hydrogen fuel every second.

WHAT IS THE MOON?

The Moon is Earth's natural satellite. A natural satellite is an object in space that orbits a planet or another body larger than itself. The biggest and brightest object in the night sky, the Moon seems to shine almost like a pale sun. It is the Earth's companion in space and circles around it continuously.

Earth's Moon

Big? WHAT ARE THE PHASES OF THE MOON?

Phases of the Moon

Only that part of the Moon lit by the Sun is bright enough for us to see. As the Moon circles the Earth, the Sun shines on it from changing angles. This is why we see more or less of the Moon at different times and the Moon appears to change shape over the course of a month as it circles Earth. These changes are called 'phases'. During the first half of the month, the Moon grows from a crescent-shaped new Moon, to a half-circle-shaped half-Moon, to a full Moon. This phase is called waxing. In the second half of the month, we gradually see less and less of the Moon, as it decreases to a half-Moon then to a crescent-shaped old Moon again. This phase is called waning.

HOW LONG HAS THE MOON BEEN AROUND?

The Moon has circled Earth for at least 4 billion years. Most scientists believe that the Moon formed when, early in Earth's history, a planet smashed into it. The impact was so great that nothing was left of the planet that hit Earth but a few fragments that were thrown back up into space. These fragments and debris from the battered Earth were drawn together by gravity and formed the Moon.

SOLAR SYSTEM

Quick-FIRE?

WHAT IS A HARVEST MOON?

The harvest moon is the full moon that occurs nearest the autumnal equinox (when night and day are of equal length).

A harvest moon

WHY DOES THE MOON LOOK THE SAME SIZE AS THE SUN FROM EARTH?

Because it is much nearer to Earth than the Sun.

ARE THERE OTHER MOONS?

As many as 240 bodies, all in our Solar System, are classified as moons.

HOW LONG IS A MONTH?

The word 'month' comes from 'Moon' and describes the approximate time it takes the Moon to go once around the Earth. This 29.53-day cycle is called a 'lunar month'. It's slightly different from the calendar months used by most people today.

Earth seen from the Moon

WHAT IS MOONLIGHT?

Moonlight

The Moon is a big, cold block of rock. It does not give out any light itself. Moonlight is simply the Sun's light reflected off the pale dust on the Moon's surface. Yet moonlight makes the Moon by far the brightest object in the night sky.

WHAT IS MERCURY'S ATMOSPHERE LIKE?

Illustration of the landscape on Mercury

Mercury is the planet nearest the Sun in our Solar System. It is 45.9 million km from the Sun at the closest point in its orbit, and 69.7 million km at its farthest. Mercury's atmosphere, also called its exosphere, is so thin that there is nothing to stop meteors smashing into its surface – and nothing to smooth out any craters. So its surface is deeply pitted with scars of meteor impacts. All you'd see on a voyage across the surface would be vast empty basins, cliffs and endless yellow dust.

Quick-FIRE?

CAN MERCURY BE SEEN WITHOUT A TELESCOPE?

Yes.

Mercury seen without a telescope

DOES MERCURY HAVE A MOON?

No.

ARE THERE MOUNTAINS AND VALLEYS ON MERCURY'S SURFACE?

Yes.

HAVE SPACECRAFT BEEN TO MERCURY?

Approaching the planet is difficult because it lies so close to the Sun. Only two uncrewed spacecraft have been close to Mercury. A third mission is underway that is due to reach the planet in 2025.

The spacecraft Messenger near Mercury

SOLAR SYSTEM | **41**

DOES **MERCURY** HAVE ICE?

Mercury has small ice caps at each pole – but the ice is frozen acid, not water!

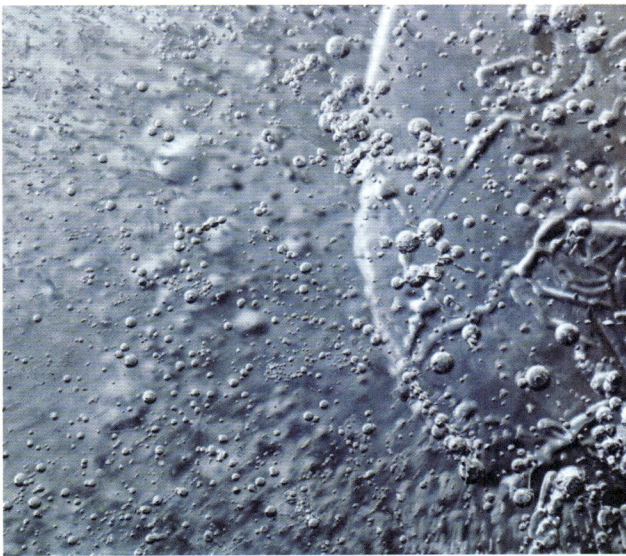

Ice caps on Mercury

The closest planet to the Sun

Big ? WHY IS THE TEMPERATURE ON **MERCURY** SO EXTREME?

Temperatures on Mercury range from one extreme to the other because the planet spins very slowly, and its atmosphere is too thin to provide any insulation. In the day, temperatures soar up to 430 °C, but at night they plunge to nearly -180 °C. This is the most extreme temperature range on any planet of our Solar System. Also, unlike Earth, Mercury barely has any tilt on its axis, so temperatures at its south and north poles are almost the same.

HOW LONG IS A DAY ON MERCURY?

A planet's day is measured by how long it takes the planet to spin once on its axis. Mercury spins slowly compared to Earth. It takes 59 Earth days to turn once on its axis.

Mercury orbiting the Sun

COULD YOU **BREATHE** ON MERCURY?

Not without your own oxygen supply. Mercury has almost no atmosphere – just a few wisps of sodium – because gases are burned off by the nearby Sun.

Mercury's inhospitable surface

WHY IS VENUS CALLED THE EVENING STAR?

Venus is the third-brightest object in the sky seen from Earth – only the Sun and the Moon are brighter. Before it was known to be a planet, people thought Venus was a star that rose in the evening. Because its position is close to the Sun, the planet can only be seen in the evening. It then disappears from the sky by midnight and is visible again just before sunrise.

Venus over the Pacific Ocean

IN WHAT WAY IS VENUS'S ROTATION UNIQUE?

Most planets in our Solar System rotate anti-clockwise on their axis, but Venus rotates clockwise. It is assumed that this change of direction came about because of a collision with an asteroid or another enormous object. As it does not tilt on its axis, Venus has no seasons.

Big? WHY IS VENUS HOT AND PINK?

Venus is the hottest planet in our Solar System, with a surface temperature of about 400 °C. Its swirling atmosphere is made up of white fumes of carbon dioxide and pink clouds of sulphuric acid. The build-up of carbon dioxide traps heat from the Sun like a greenhouse, which is why Venus is so incredibly hot. The atmosphere, made of the fumes of gases from volcanoes, is so dense that the pressure it exerts on the exterior of the planet is 90 times greater than that on Earth and its surface cannot be seen at all.

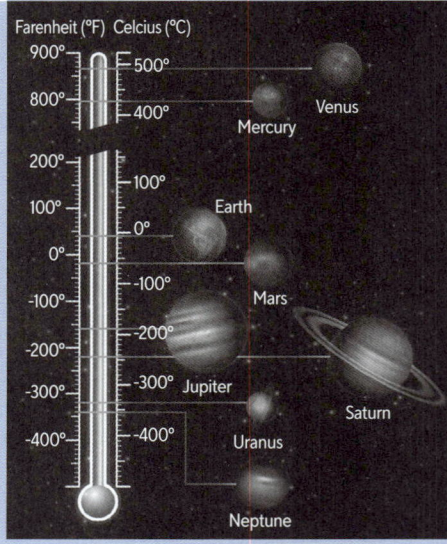

The temperature of the planets

WHAT IS INSIDE VENUS?

The cloud cover and extreme pressure make it difficult to gather a lot of information on Venus. It is thought that, much like Earth, Venus has three layers – crust, mantle and core. It does not have a magnetic field so scientists assume that the core is solid.

Looking inside Venus

WHY IS VENUS CALLED EARTH'S SISTER PLANET?

A similarity in size and mass results in Venus frequently being referred to as Earth's sister planet. The difference in diameter is only 638 km. The planets lie closest to each other in the Solar System, with the average distance between them being 61 million km.

The sister planets

Quick-FIRE?

DOES VENUS SHOW TRACES OF WATER?

Yes, there are traces of water on Venus.

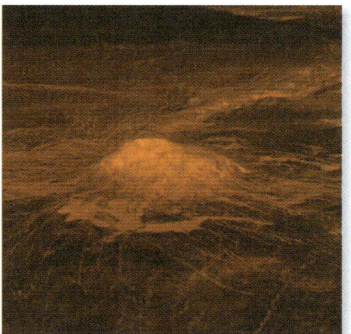

Venus's surface

HOW DO WE STUDY THE SURFACE OF VENUS?

Through radar mapping, which creates images of the landscape.

Maat Mons volcano

HOW MANY MOONS DOES VENUS HAVE?

None at all.

WHO IS VENUS NAMED AFTER?

The Roman goddess of love and beauty.

HOW LONG IS A DAY ON VENUS?

Venus turns very slowly on its axis; so slowly in fact, that it manages to go around the Sun in 225 days, whereas it takes 243 days to complete a single turn on its axis. This means that on Venus a day is longer than a year!

Slow-moving Venus

WHY IS MARS CALLED THE RED PLANET?

Mars, the fourth planet from the Sun, has a large amount of iron dust covering its surface. Oxygen in the planet's atmosphere reacts with this iron (known as oxidisation), giving it a rusty-red colour. During the huge dust storms that Mars is prone to, the planet can be wrapped in red dust for months, earning it the title 'Red Planet'.

The surface of the 'Red Planet'

Phobos and Deimos

WHO DISCOVERED MARS'S MOONS?

In 1877, American astronomer Asaph Hall discovered Mars's two moons. He named them Phobos and Deimos, after the attendants of the Roman war god, Mars. These two moons are among the smallest in the Solar System.

Big? WHAT IS THE SURFACE OF MARS LIKE?

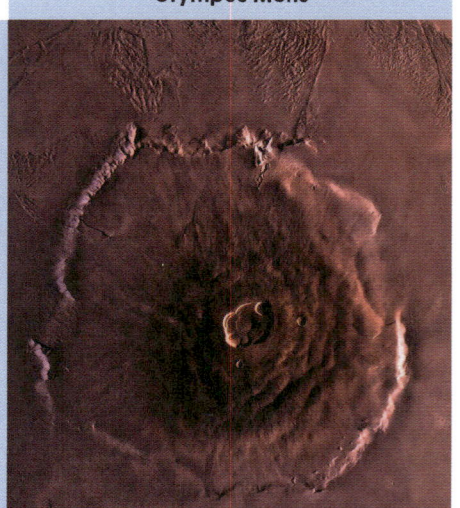
Olympus Mons

Mars's atmosphere is thinner than Earth's, with more than 95 per cent carbon dioxide and less than one per cent oxygen. The average temperature is way below freezing. The surface of the planet is rocky, with canyons, volcanoes, dry lake beds and craters. The volcanoes are extinct, but Olympus Mons, three times taller than Mount Everest, is the biggest volcano in the Solar System. Valles Marineris, a 4,000-km-long canyon that is four times as deep as the Grand Canyon, is the largest known gorge in the Solar System.

SOLAR SYSTEM | 45

IS THERE LIFE ON MARS?

In the 1970s, uncrewed landers from the Viking missions found no trace of life. In 1996 microscopic fossils of what might have been mini viruses were seen in a rock from Mars. However, at the moment, scientists agree that no positive signs of life have been found.

No signs of life

WHAT DO THE MARKINGS ON MARS SAY?

Seasonal markings, dark finger-like streaks that look like rivers, streams and small channels, appear along the hills and slopes of Mars, swelling during the planet's warm season and fading as it gets cold. Although they were once thought to be signs of civilisation, they are purely geological.

WILL HUMANS EVER LAND ON MARS?

It is very probable that they will, but a crewed mission to Mars is unlikely before the late 2030s. The safety of the astronauts will be a main concern. Among other challenges, Mars does not have an ozone layer that protects humans against the Sun's ultraviolet radiation.

Uncrewed exploration of Mars

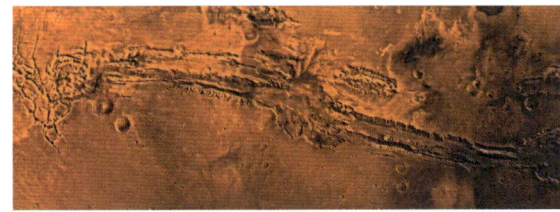
Intriguing formations on the surface of Mars

Quick-FIRE?

The Mars lander carried by the Pathfinder

WHICH SPACE PROBE LANDED ON MARS IN 1997?
The Mars Pathfinder.

Sunset on Mars

HOW FAR IS MARS FROM THE SUN?
Mars is nearly 227.9 million km from the Sun.

The rusty-red planet glows in the sun

HOW BIG IS MARS?
Mars's diameter is nearly 6,792 km. Its mass is just over one-tenth of Earth's.

A barren landscape

ARE THERE ANCIENT PYRAMIDS ON MARS?
No. There are no visible signs of civilisation.

HOW BIG IS JUPITER?

The fifth planet from the Sun, Jupiter is a giant. It is twice as heavy as all the other planets put together. Though made up mostly of gas, it is so big that 1,300 planets the size of Earth could fit inside it. Its diameter of approximately 139,820 km is more than 11 times that of Earth.

Jupiter, with its ring and stripes, and a few of its moons

HOW FAST DOES JUPITER SPIN?

The fastest spinner

Despite its huge size, Jupiter spins faster than any other planet. It turns once round in just 9 hours 55 minutes, which means that the surface is moving at an astounding 45,000 km per hour.

IS JUPITER LIKE A BIG MAGNET?

Jupiter's bulk and the speed at which it spins churn the metal insides of the planet so much that the planet becomes a giant generator, creating a magnetic field that is 14 times as strong as Earth's.

IS JUPITER FULL OF GAS?

Swirling clouds on the surface of the planet

Jupiter is a gas giant made up mostly of hydrogen and helium. But the planet's gravity is so high and the internal pressure so great that the gases turn to liquid. Beneath Jupiter's thin atmosphere of ammonia, there is a 25,000-km-deep ocean of liquid hydrogen. The planet is covered in thick clouds, which look like stripes. These clouds have a temperature of about -145 °C. The temperature near the core is much higher, approximately 24,000 °C.

SOLAR SYSTEM | **47**

Quick-FIRE ?

WHAT GENERATES JUPITER'S STORMS?

The energy for the storms does not come from the Sun but is radiated by Jupiter itself.

A stormy planet

HOW MANY VOLCANOES DOES JUPITER'S MOON, IO, HAVE?

More than 400! About 150 of these are active.

Io

HOW MUCH TIME DOES JUPITER TAKE TO ORBIT THE SUN?

About 12 Earth years.

WHAT IS JUPITER'S RED SPOT?

The Great Red Spot, or GRS, is a huge swirling storm in Jupiter's atmosphere that has been raging for at least 330 years. Its size is much wider than the diameter of Earth, though it seems to have been shrinking over the years. Spacecraft have taken vivid images of the storm with its red centre and wavy cloud formations.

The Great Red Spot marks a storm

COULD YOU LAND ON JUPITER?

No. Even if a spaceship could withstand the enormous pressure, Jupiter has no surface to land on. The atmosphere merges directly with the deep oceans of liquid hydrogen.

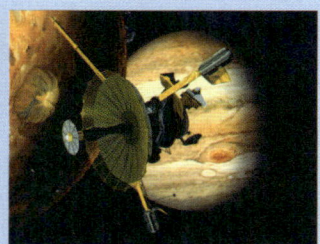

Rendering of the spacecraft Galileo flying past Jupiter

HOW MANY MOONS DOES JUPITER HAVE?

Jupiter has 80 confirmed moons. Italian astronomer Galileo discovered the four largest – Io, Europa, Ganymede and Callisto – in 1610, which is why they are called the Galilean moons. Two of Jupiter's largest moons, Io and Europa, which are about the size of Earth's Moon, can be seen in orbit.

WHAT ARE SATURN'S RINGS?

First seen by Italian astronomer Galileo Galilei in 1610, through his simple telescope, Saturn's rings are the planet's shining halo. The rings are made of billions of chips of ice and dust. A few are bigger than a refrigerator and most are smaller than an ice cube. The rings are very thin, no more than 50 m deep, and stretch approximately 282,000 km into space.

Saturn with its shining halo of rings

Big? WHAT SETS SATURN APART?

Saturn is the most beautiful planet in the Solar System, with more moons than any other. The gas giant is the sixth planet from the Sun, and orbits it in 10,759 days. It has a diameter of 120,536 km and a mass of 600 billion trillion tonnes, which is over 95 times that of Earth. Winds gust across its surface at about 1,800 km per hour, faster even than those on Jupiter. Made up mostly of hydrogen and helium, the planet is surprisingly light – in a bathtub big enough to hold it, Saturn would float.

WHAT IS THE CASSINI DIVISION?

Saturn's rings occur in broad bands, referred to by the letters A to G. In 1675, the astronomer Cassini spotted a dark gap between rings A and B. This is now named the Cassini Division, after him.

The Cassini Division

SOLAR SYSTEM | 49

COULD THERE BE LIFE ON SATURN?

No. Saturn's atmosphere does not allow the existence of life. The temperature and pressure are both extreme and it is too changeable an environment to support life.

WHY IS TITAN SPECIAL?

Titan, one of Saturn's many moons, is very special as it is the only moon in the Solar System with the kind of dense atmosphere that is vital for supporting life.

Artist's rendering of one of Saturn's moons

HOW OLD IS SATURN?

It is estimated that Saturn was formed along with the rest of the Solar System about 4.5 billion years ago, when gravity pulled in swirling masses of gas and dust to become this huge gas giant.

Saturn dwarfs Earth

Quick-FIRE?

WHICH OF SATURN'S MOONS WAS ALMOST SPLIT IN HALF BY A GIANT METEOR?

Tethys.

Tethys

WHICH OF SATURN'S MOONS IS BLACK AND WHITE?

Iapetus is black on one side and white on the other.

WHAT IS THE TEMPERATURE ON SATURN?

The average temperature in its upper atmosphere is around -175 °C.

HOW DID SATURN'S RINGS FORM?

Looking closely at the rings

Saturn's rings are assumed to be pieces of comets, asteroids or shattered moons that may have broken up before they reached the planet. Saturn's powerful gravity further shattered them into billions of small chunks of ice and rock coated with dust.

WHEN WAS URANUS DISCOVERED?

Uranus is the seventh planet from the Sun in our Solar System. To ancient astronomers, it appeared as a faint star as it is very far away from Earth. When British astronomer William Herschel first discovered it in 1781, he initially thought it was a comet but soon realised his mistake.

Uranus is tilted over on its side

DOES IT RAIN DIAMONDS ON URANUS?

Yes! The extreme pressure deep in Uranus's atmosphere combines carbon and hydrogen to form diamonds, which then sink down to settle around the core of the planet.

HOW IS URANUS DIFFERENT FROM OTHER PLANETS? *Big*

Unlike other planets, Uranus does not spin at a slight angle. Instead, it is tilted right over and rolls on its side in the opposite direction from Earth and most other planets. Like the other gas giants, the atmosphere is made up of hydrogen and helium, but beneath flow oceans of liquid methane. The planet is very cold, and as winds of over 2,000 km per hour whistle through its atmosphere, they create huge waves in the icy oceans of methane, giving the planet its beautiful, blue-green colour.

Inside Uranus

- Outer atmosphere, the upper cloud layer
- Atmosphere (hydrogen, helium, methane gases)
- Mantle (water, ammonia, methane ices)
- Core (Silicate/Fe-Ni rock)

IS URANUS THE COLDEST PLANET IN OUR SOLAR SYSTEM?

Though Uranus is not the farthest planet from the Sun in our Solar System, it is the coldest. Its core is so cold that it does not radiate much energy. This means the planet gives off less heat than it receives from the Sun and the surface temperature drops as low as -224 °C.

Uranus's place in the Solar System

Quick-FIRE?

HOW LONG IS A YEAR ON URANUS?

One year on Uranus equals 84 Earth years.

Earth and Uranus

HOW LONG ARE SUMMER AND WINTER ON URANUS?

They each last 42 years.

CAN URANUS BE SEEN WITH THE NAKED EYE?

Yes, in an extremely dark sky.

HOW MANY MOONS DOES URANUS HAVE?

The existence of 27 moons has been confirmed. Most are small and irregular, and made of ice and rock. Each of the largest moons, including Miranda, are roughly presumed to be made of equal parts water ice and silicate rock. Ariel is the brightest moon.

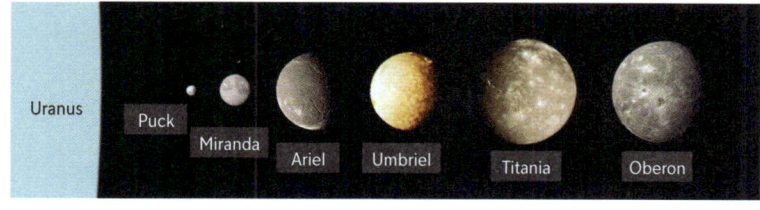

Some of Uranus's moons

HOW DID URANUS GET ITS NAME?

Uranus was named after the Greek sky god Ouranos. Many names were proposed, but this was the name finally selected by German astronomer Johann Bode who mapped the orbit of Uranus.

DOES URANUS HAVE RINGS?

Uranus has rings, like the other gas giants. So far 13 have been identified, of which the Epsilon ring is the brightest. Uranus's rings are smaller than those of Saturn and are made of extremely dark particles, which are no more than a micrometre to a fraction of a metre in size and, therefore, not very noticeable.

Uranus has many rings

HOW WAS NEPTUNE DISCOVERED?

When studying Uranus, astronomers were intrigued by a disturbance in its orbit that seemed to be the gravity of another planet pulling at it. But there was no such planet visible to the naked eye. Then, in the early 1840s, mathematicians John Couch Adams in England and Urbain Le Verrier in France calculated where the pull was coming from. On 23 September 1846 Johann Galle in Berlin spotted the mystery planet, which was named Neptune.

Dark stormy spot on Neptune's surface

Neptune's distinctive colour

Big? IS NEPTUNE THE WINDIEST PLANET?

Yes, Neptune does have the strongest winds known in our Solar System – its record wind speed has been measured at about 2,400 km per hour. In 1989 the largest storm ever was recorded with high-speed winds swirling around the planet in a dramatic and extraordinary manner. The Great Dark Spot, as it was called, lasted about five years. Scientists are still trying to work out how a planet so far from the Sun generates such powerful winds.

HOW LONG IS A YEAR ON NEPTUNE?

One orbit around the Sun takes Neptune 164.81 Earth years. Neptune is about 4,459 million km from the Sun at its closest, and about 4,536 million km at its farthest.

WHAT COLOUR IS NEPTUNE?

The atmosphere on Neptune has methane gas in it. The gas absorbs all the red light from the colour spectrum and gives Neptune a vivid, azure, blue-green appearance.

Methane gas colours Neptune blue

SOLAR SYSTEM

HOW MANY MOONS DOES NEPTUNE HAVE?

Neptune has 14 confirmed moons, the largest of which is Triton. William Lassell, an amateur astronomer, discovered Triton on 10 October 1846, just 17 days after the planet was first discovered.

IS NEPTUNE THE SMALLEST OF THE GAS GIANTS?

Yes, Neptune's diameter is 49,528 km. However, it is much denser than the other gas giants. It weighs 17 times more than Earth, while Uranus is only 14 times heavier than Earth.

The smallest gas giant

Rings that are not easily seen

DOES NEPTUNE HAVE RINGS?

Yes, Neptune has at least five main rings, probably made of ice and dust particles. However, they are not easy to see, as they are very thin. The planet also has four ring arcs that are clumps of dust and debris scattered from nearby moons.

Quick-FIRE?

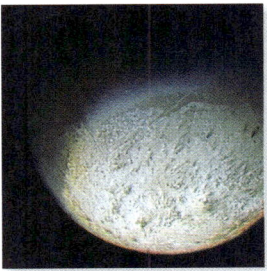

Triton

WHAT IS PECULIAR ABOUT TRITON?

It orbits backwards in relation to the other moons of Neptune.

The Roman god Neptune

WHO IS NEPTUNE NAMED AFTER?

The Roman god of the sea.

IS TRITON THE COLDEST PLACE IN OUR SOLAR SYSTEM?

No! At about -235 °C, it was once believed to be the coldest. However, we now know that the Hermite Crater on Earth's moon is an icy -248 °C.

WHY IS PLUTO NOT A PLANET?

Initially, our Solar System was considered to have one Sun and nine planets. It was in 2006 that the International Astronomical Union (IAU) reclassified Pluto as a 'dwarf planet'. Although Pluto, like a planet, is a sphere and orbits the Sun, it did not qualify as a planet because other similar-sized objects share its gravitational space.

Pluto and its moon Charon

Quick-FIRE ?

The Roman god Pluto

WHO IS PLUTO NAMED AFTER?
The Roman god of the underworld.

Earth, Earth's Moon and Pluto

HOW MANY EARTH DAYS EQUAL ONE DAY ON PLUTO?
6.39 Earth days.

HOW LONG IS A YEAR ON PLUTO?
248.89 Earth years.

IS PLUTO VERY COLD?
Yes, about -223 °C.

WHAT IS PLUTO MADE OF?

It is believed that, apart from its rocky core, Pluto is made of nitrogen ice, methane and carbon monoxide. Its surface, a range of colours from charcoal black to dark orange and white, has a mountain made of water-ice and a heart-shaped glacier.

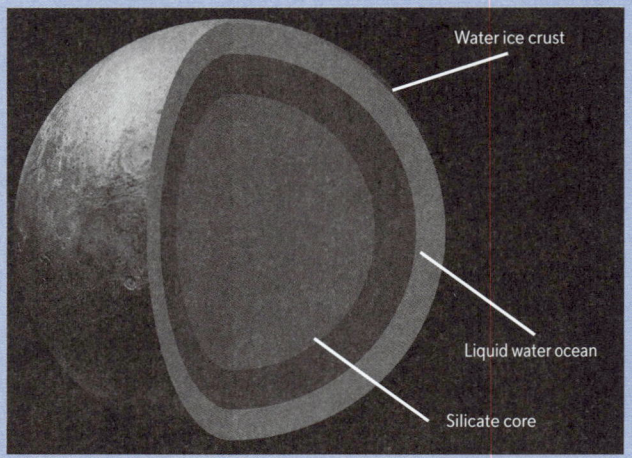

Pluto's internal structure

SOLAR SYSTEM | **55**

WHAT ARE DWARF PLANETS?

Including Pluto, there are currently five known dwarf planets in the Solar System. The other four are Ceres, Makemake, Eris and Haumea. These bodies are large enough to be pulled into a sphere-shape by their own gravity, but their gravity is not strong enough to clear the area around them of other bodies.

(Left to right) The dwarf planets Pluto, Eris, Haumea, Makemake and Ceres

Big? WHAT IS FASCINATING ABOUT PLUTO'S MOONS?

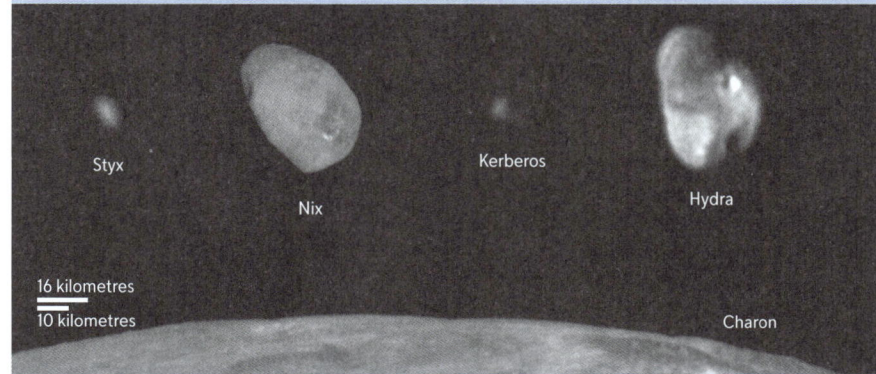

Charon, and Pluto's other, smaller moons

Despite being very small, Pluto has a complex collection of satellites. It is believed that the dwarf planet's moons were formed by a collision between two other dwarf planets and a Kuiper Belt object, when the Solar System was still young and in the process of being formed. There are five confirmed moons – Charon, Nix, Hydra, Kerberos and Styx. Pluto is only twice as big as Charon. Both orbit each other and are locked together in what is sometimes called a double dwarf planet system.

HOW SMALL IS PLUTO COMPARED TO THE REST OF THE SOLAR SYSTEM?

Pluto is not only smaller than the planets but also many moons in our Solar System, such as Ganymede, Titan, Callisto, Io, Europa and Triton. Its diameter is 66 per cent of that of Earth's Moon and it has only 18 per cent of its mass.

Pluto's size as compared to some moons in the Solar System

WHAT MAKES EARTH A TERRESTRIAL PLANET?

The third-farthest from the Sun in our Solar System, Earth is the fifth-largest planet in size and mass. Its interior is mainly rocks and metals, which is why it is called terrestrial. One-third of its surface is made up of silicate rock divided into plates, and the remaining two-thirds are covered by oceans – all of which makes it unique among planets.

COULD EARTH HAVE A TWIN?

Out in the Universe beyond our Solar System, Kepler-452b orbits a star similar to the Sun. Though not Earth's identical twin – Kepler-452b is slightly bigger – it is remarkably similar. Its orbital path is at about the same distance as Earth's orbit of the Sun, and it is quite likely to have an atmosphere and water as well.

The 'twins' placed side by side

Looking at Earth from space

Quick-FIRE?

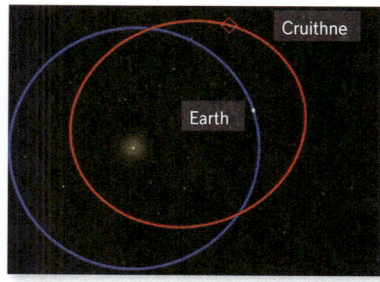
The orbit of 3753 Cruithne

WHAT IS 3753 CRUITHNE?
An asteroid-type object that orbits both the Sun and Earth.

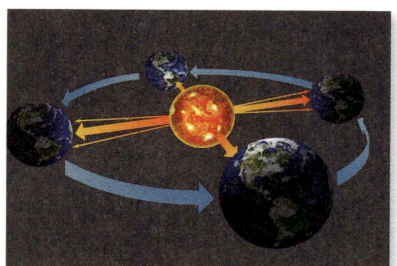
Earth's orbit around the Sun

AT WHAT SPEED DOES EARTH ORBIT THE SUN?
30 km per second.

Earth's widest point is the Equator.

WHAT IS EARTH'S DIAMETER?
Approximately 12,800 km.

HOW HOT IS EARTH'S INNER CORE?
About as hot as the surface of the Sun.

WHAT IS SPECIAL ABOUT EARTH'S SIZE AND DISTANCE FROM THE SUN?

Earth is just the right size to hold on to its atmosphere. Its distance from the Sun is also just right. Had it been any closer, the heat would have caused all its surface water to evaporate; any farther and it would have frozen.

Earth is placed just right

WHY IS EARTH THE DENSEST BODY IN THE SOLAR SYSTEM?

Earth's dense core

Because Earth has a dense composition of different elements and compounds. Also, the high force from gravity on Earth significantly compresses this dense mass. Comparatively, the Sun's density is just a quarter of Earth's.

HOW MANY NATURAL SATELLITES DOES EARTH HAVE?

In addition to the Moon, Earth could have two additional 'satellites' that are made up of dust. Called the 'Kordylewski clouds' after the Polish astronomer who spotted them, they are about the same distance away from Earth as the Moon, but are seen as merely a faint glow in the sky.

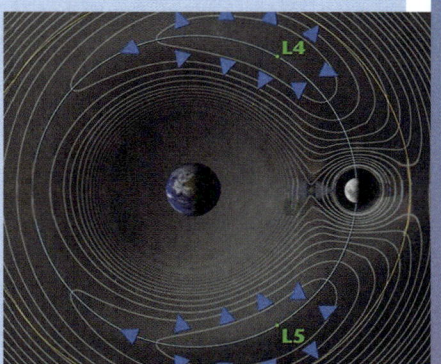

L4 and L5 mark the Kordylewski clouds

Big? HAS EARTH'S ROTATION SLOWED DOWN?

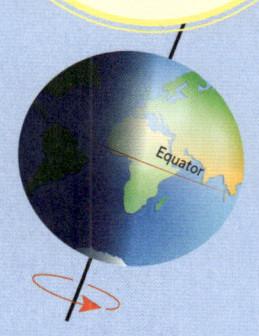

Earth rotates on a tilted axis.

The circular movement of Earth on its own axis is referred to as rotation. This movement is anti-clockwise and it takes 23 hours 56 minutes and 4 seconds – that is, almost 24 hours – for Earth to complete one rotation.
Each rotation marks a day. Its rotational speed is slowing down due to the effect of the Moon on tides, but this is happening very slowly.

WHAT IS AN ASTEROID?

British Astronomer William Herschel coined the name 'asteroid' in 1802. This name identified the thousands of lumps, mostly made of rock and metal, that circle the Sun in a big band between Mars and Jupiter – though some do venture outside this zone.

WHAT ARE PERIODICS?

Some comets appear at regular intervals, which is why they are called 'periodics'. Encke's Comet passes by Earth every 1,206 days; Halley's Comet every 76 years.

WHAT IS A METEOR SHOWER?

Meteor shower at night

Every day, tonnes and tonnes of space debris rain down on the Earth, sometimes in such great concentrations that they make a golden rain in the night sky as they hit the Earth's atmosphere and burn up. This is called a meteor shower. Meteors are space dust and lumps so small that they burn up long before they hit the ground.

Quick-FIRE ?

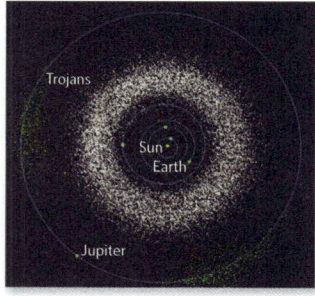
The Trojans

WHAT ARE TROJAN ASTEROIDS?
A small group of asteroids that circle the Sun on the same orbit as Jupiter.

Meteor crater in Arizona, USA

HOW MANY METEORITES HAVE BEEN DISCOVERED?
More than 50,000, some as small as marbles.

Edmund Halley

WHO DISCOVERED HALLEY'S COMET?
Edmund Halley.

WHAT IS A COMET'S TAIL MADE OF?
Dust and gases blown backwards by the solar wind.

OUR PLANET | 59

A fragment of the Canyon Diablo meteorite

Halley's comet

Big? WHAT IS A COMET?

Comets are dirty ice-balls. They circle the Sun like the planets, but have very long, stretched-out orbits and spend most of their time in the far reaches of the Solar System. Occasionally, when one of them is drawn in towards the Sun, it becomes visible for a few weeks. As a comet hurtles along, material from its surface is blown backwards and stretches out behind it. This 'tail' can be seen in the night sky, shining spectacularly as it catches the light of the Sun until it swings out of sight.

HOW FAST IS A COMET?

Comets speed up as they near the Sun, reaching up to 2 million km per hour. But away from the Sun, they slow down to speeds of little more than 1,000 km per hour, which is why they stay away for so long.

Comet Hale-Bopp

WHICH ARE THE BIGGEST AND BRIGHTEST ASTEROIDS DISCOVERED?

Vesta and Earth's Moon

The biggest and brightest asteroid is called Vesta. It measures at 530 km across and is the only asteroid that can be seen with the naked eye.

HOW ARE ASTEROIDS NAMED?

Asteroids were first discovered in 1801 by a group of astronomers, calling themselves the Celestial Police, hunting for a missing planet they were certain lay somewhere between Mars and Jupiter. Nowadays, new asteroids are discovered frequently. Each is given an identification number and is named by the discoverer. Names vary – some are even named after Greek goddesses.

Naming asteroids

WHAT IS AN ASTRONAUT?

A person specially trained to explore and study space, and who travels in a spacecraft. Russians use the word 'cosmonaut' instead. Astronauts go through strict training in every aspect of space travel, gaining technical knowledge and developing high fitness levels so that they can travel beyond Earth's gravity and remain healthy in a weightless environment during their journey.

Astronaut taking a spacewalk

DO ASTRONAUTS HAVE TO BE CAREFUL ABOUT THEIR HEALTH?

Yes, the extreme differences in conditions on Earth and in space could cause bones to weaken, muscles to shrink, injuries due to radiation, sleep disturbances and other health problems. That is why physical fitness is highly important for an astronaut.

Health check-up in space

Big? WHY DO ASTRONAUTS WEAR SPACESUITS?

Spacesuits are special garments designed to protect astronauts from the harsh, extreme environment of outer space. Astronauts usually wear a spacesuit when stepping outside the spacecraft for exploration and study. The suit helps protect them from the Sun's radiation and other extremes. In case of sudden loss of cabin pressure, or other emergency situations, they can even wear it inside the spacecraft.

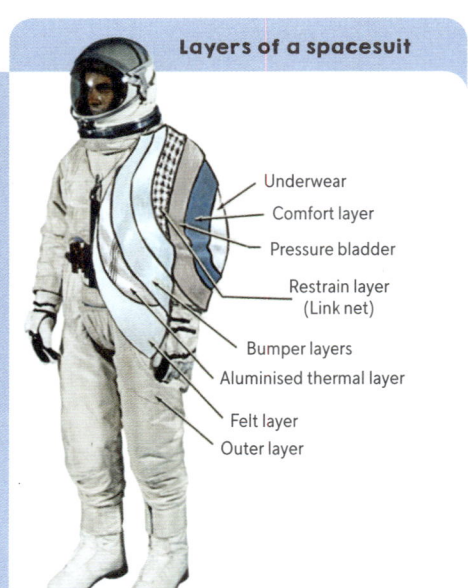
Layers of a spacesuit
- Underwear
- Comfort layer
- Pressure bladder
- Restrain layer (Link net)
- Bumper layers
- Aluminised thermal layer
- Felt layer
- Outer layer

OUR PLANET | **61**

WHAT DO ASTRONAUTS EAT AND DRINK IN SPACE?

Astronauts carry food packages that give them the correct nutrients during their time in space. There are various foods astronauts can eat: fruits, nuts, meat, breads, and sweets are some of them. Salt and pepper are available in liquid form. Drinks like tea, coffee and juices can also be consumed.

Eating floating burgers in space

HOW MANY COUNTRIES HAVE SENT CITIZENS INTO SPACE?

Astronauts and cosmonauts

The Soviet Union was the first country to send a human to space in 1961. Since then 41 countries have sent people into space.

Quick-FIRE ?

CAN A SPACE SHUTTLE FLY TO THE MOON?

No, it is designed to hover around Earth's orbit.

Columbia shuttle

WHAT IS THE ISS?

The International Space Station, specially constructed as a multinational project to conduct missions and experiments in space.

ISS

WHO MAKES UP A SPACE EXPLORATION TEAM?

The team consists of a commander, pilot and mission specialists.

CAN ASTRONAUTS TAKE A SHOWER IN SPACE?

No, they take only sponge baths.

WHAT IS THE ORIGIN OF THE NAME 'ASTRONAUT'?

The word 'astronaut' is derived from the Greek *astron nautes* meaning 'star sailor'. The term has been in use since the 19th century in various short stories and novels about space and exploration.

HOW DID LIFE BEGIN?

Scientific experiments have shown how lightning flashes could create amino acids, the basic chemicals of life, from the waters and gases found on an early Earth. But no one knows how these chemicals were able to make copies of themselves. This is the key to life, which remains a mystery. Another theory is that life could have come to Earth on a comet or asteroid from a distant part of the Universe.

How life began remains a mystery.

Big? IS IT POSSIBLE THAT EXTRATERRESTRIAL LIFE CAME TO EARTH FROM MARS?

Meteorite ALH84001

In 1984 a meteorite from Mars was found by geologists in Allan Hills, Antarctica. Study of the meteorite, called ALH84001, revealed tiny crystals similar to those formed by bacteria on Earth. Also discovered were worm-like patterns that could have been fossils. However, it was later agreed by most scientists that these crystals could also form without the help of bacteria, and the 'fossil patterns' were quite likely created while the rock was being investigated in the lab.

WHAT IS SETI?

The Search for Extraterrestrial Intelligence (SETI) is a scientific effort to continually scan radio signals from space with the aim of picking up signs of intelligence. SETI scientists look for signals that have a pattern, but are not completely regular, like those from pulsating stars. Such signals could be sent by alien civilisations with or without the intention of making contact with other civilisations like ours.

SETI Institute, California, USA

OUR PLANET | 63

Launching the space telescope

WHAT WAS THE KEPLER MISSION?

It was a space telescope launched in 2009 to discover Earth-like planets revolving around other stars that could hold life. Kepler used the transit method – the dimming of a star's light when a planet crossed in front of it – to discover more than 2,600 new planets. Many of these planets were in the 'sweet spot' – the right distance from the parent star to support life.

Quick-FIRE?

WHAT WERE SPIRIT AND OPPORTUNITY?

Two rovers that landed on Mars to look for the presence of water, which could be a sign that life once existed on the planet.

Mars rover

WHAT IS THE GOLDEN RECORD?

A phonograph disc, with sounds and images from Earth, carried out of our Solar System by the Voyager spacecraft.

The Golden Record sent into space

ARE RADIO SIGNALS THE ONLY WAY TO DETECT EXTRATERRESTRIAL LIFE?

No, laser pulses and evidence of gas pollutants could also be signs.

WHAT IS LIFE MADE OF?

Life is based on compounds of the element carbon, known as organic chemicals. Carbon compounds called amino acids link up to form proteins, and proteins form the chemicals that build and maintain living cells.